The Wildwood

Kingley Vale: detail of a yew trunk.

The Wildwood

In Search of Britain's Ancient Forests

Text by Gareth Lovett Jones
and Richard Mabey

Photographs by Gareth Lovett Jones

AURUM PRESS

First published 1993 by Aurum Press Ltd, 10 Museum Street, London WC1A 1JS
Text copyright © 1993 by Richard Mabey and Gareth Lovett Jones
Photographs copyright © 1993 by Gareth Lovett Jones

A catalogue record for this book is available from the British Library

ISBN 1 85410 242 7

2 4 6 8 10 9 7 5 3 1
1994 1996 1997 1995 1993

Designed by Robert Updegraff
Typeset by Wyvern Typesetting Ltd
Printed in Singapore

Contents

THE WILDWOOD 7

by Richard Mabey

✳

Glen Strathfarrar: the forest floor.

The Wildwood

– BY RICHARD MABEY –

Oliver Rackham, doyen of landscape historians, has written that:

> The landscape is like a historic library of 50,000 books. Many were written in remote antiquity
> in languages which have only lately been deciphered; some of the languages are still unknown . . .
> Every year a thousand volumes are taken at random by people who cannot read them, and sold
> for the value of the parchment.[1]

In this metaphor ancient woodlands are the most densely written and ornamented manuscripts of
all. There is nothing in the landscape that can match their intricacy of structure or the extravagance
of natural life that has found a home in them. They are places where individual trees may survive
for more than a thousand years; where orchids flower in deepest shade; where there are cycles of
decay, regrowth and mutual sharing of awesome intricacy. They are the incunabula of the natural
world, unique and irreplaceable.

Yet they continue to be destroyed, and not just in remote corners of the globe. It has been an
uncomfortable experience for many Europeans, used to regarding deforestation as a problem of the
tropics, to be reminded of how early and how thoroughly we cleared our own natural forests. Even
in Britain, we were still destroying ancient woods into the mid-1980s, until we were down to our
last 300,000 hectares.

No wonder those that remain have begun to be regarded with a new kind of respect. They are
not just 'reserves' in the conventional sense, but Arks, life-rafts out of the past, and in them we can
believe ourselves to be in touch with something unsullied by civilization, with a wellspring of
nature. It is an understandable longing, and only partly a futile one.

The remarkable portraits of Britain's ancient woodland reproduced in this book came out of just
such a quest for Gareth Lovett Jones. One question dogged him repeatedly:

> What might it have been like to stand . . . or crawl, or force one's way onward through the
> primeval forests? Would it be possible to build up some kind of an impression – albeit a fanciful
> and, it need hardly be added, a wholly unscientific impression – of such an experience by
> penetrating some of the tiny surviving fragments of wooded wilderness or quasi-wilderness, and
> photographing them as if man had never been there?

In a purist sense, of course, there is no such possibility. There are no fragments of 'virgin' forest in Britain, no stretch of pristine woodland that has been entirely free from human influence. Even woods in impenetrable gorges that may never have been cut or walked in are not immune to acid-rain and man-made drought. They have suffered too from changes in the woodland beyond them, with which they might once have had a free trade in plants and creatures.

The terms used to describe them are similarly compromised and relative. 'Ancient woodland' is used conventionally for any woodland (usually unplanted) that dates from before 1600. 'The wildwood' was a term originally coined by Oliver Rackham to denote Britain's natural, aboriginal forest cover, but it is now commonly used for any remaining – albeit managed – fragments of this forest, as well as more vaguely for any large patch of spontaneous, feral woodland. If the terms are indefinite, they do at least reflect the muddled state (and origins) of most existing woods. And 'the wildwood' itself reflects Kenneth Grahame's 'Wild Wood' in *The Wind in the Willows*, that menacing wilderness of pitfalls and hollow beeches where Mr Badger – and the stoats and weasels – lived. The idea of this expanse of forest, 'low and threatening like a black reef in some still southern sky', separating the enchanted world of the river bank from the Wide World beyond, has had a powerful influence on our feelings about ancient woodland. Yet Grahame makes it clear that the Wild Wood is not really ancient at all, but had 'planted itself and grown up' on the site of an old city. To underline the ambivalence still further, Grahame based his Wild Wood on a real and authentically ancient tract, the steep billow of Bisham Woods, which stretched along the Chiltern escarpment near his home in Cookham Dean and which have been there since the Ice Age, despite cropping, coniferization and storms.[2]

Yet the *sense* of wildness (and sometimes even of wilderness) in Britain's ancient woods is not entirely an illusion. They are not landscape fossils, collections of wooden standing stones. They are alive, and evolve, adapt, decay and regenerate in ways that we do not fully control or even fully grasp. They can absorb and to some extent 'naturalize' human impact. And more, perhaps, than any other kind of natural community, they challenge the distinctions we set up between change and persistence. Their special relationship with time makes them, as Gareth Lovett Jones's work demonstrates, apt subjects for respectful photography. An ancient woodland is physically massive and cryptic, and its rhythms of change minutely slow. At any moment it is a kind of 'still' of its own life, beyond reduction or simplification. It will resist attempts to pose or 'capture' it, always hold something back. Picturing it is just the beginning of an exploration, not an end.

I have sensed this quest, this search for the quintessential greenwood, going on in my own life. I grew up in wooded country, spent much of my childhood up trees, and now own a fragment of ancient woodland myself.[3] It is in the Chilterns, not far from several towns and villages, and I have tried to run it as a kind of community patch with – and for – the people in the neighbourhood. We clear ponds, thin trees and make space for the more diffident plants. We make space for ourselves, too, as a tribal species that rather enjoys playing in woods. The wood consequently has a social as well as an ecological life, and each May there is a ritual there that seems to catch the complex feelings we have about these ancient, living monuments.

In the middle of the month, on Ascension Day, the children from the village school walk across the fields and hold a service amongst the bluebells. Under the opening beech leaves they sing hymns to new life and the mysteries of transubstantiation. I am not religious myself, but I sometimes join the adults on the sidelines and, lumps in throats, we gaze heavenwards too, and give quiet thanks that the woodland canopy is opening again without a sign of acid-rain damage. We are all too aware that, in the hour or so it takes to complete the service, a slice of tropical forest as big as the whole parish will have been flattened, along with all its plants and birds of paradise and intricate history of human use. This 7-hectare wood is a barely significant fragment by comparison, but on a day such as this it feels like a last bastion, a refuge for vanishing values as well as threatened creatures.

The sense of poignancy does not last. After the service the children play furious games between the trees, the green tangle swells visibly around us, and for a moment we revel in its sheer natural exuberance. Tomorrow the only reminder that we were here will be a thin criss-cross of trails through the bluebells, moist with sap.

Ceremonies like this, acted out amongst immemorial woodland, have been held as celebrations of fertility since well before the birth of Christ, and must always have expressed these complementary reverences for new life and ancient rootedness. They celebrated wildness and vitality on the one hand, and quiet retreat on the other; the energy that turns sunlight and water into living tissue, and the stolidity that enables trees to become resilient landmarks, outliving whole dynasties of humans. [...]selves – fugitive greenleaf on long-standing trunk – have always been evidence [...]thing contradictory in these two images.

[...] this wood, and in a strictly legal sense this is true. But I have never been able [...] idea of woods as private property. More than any other kind of landscape they [...], with generations of shared natural and human history inscribed in their struc- [...]k along the main track, following the path taken by the children, I can cross the [...] of 8,000 years of woodland evolution in less than 100 yards. The track itself is [...]n, cut just ten years ago, a sinuous, sloping gully of clay and crushed chalk, lying [...]h the evening sun. It tilts down from a 100-year-old beech plantation, crosses a [...] and ditch, dug to keep cattle out of growing coppice, and at the bottom joins the [...] much older boundary dike, excavated perhaps by Celtic tribes. Along the track there [...]f scarce and exquisite woodland plants that may have been growing in this place for [...] drifts of ferns, clumps of intensely parochial grass species, bowers of climbing vetches. [...] track's rim, though already being obscured by miniature landslips and bristling tufts of [...]joy and raspberries, is a meagre layer of brown forest soil, no more than nine inches [...]es directly on top of the local flinty clay, a sign that the earth has never been ploughed or [...] disturbed and that there has probably always been a wood on this site.

[...]e is no such clear boundary between the natural and the man-made here, and this idea – that [...]rdinary woods may have been flourishing for perhaps 10,000 years – is something we have [...]st begun to comprehend. One of the many myths we have inherited about the history and [...]ings' of woods is that all modest, tidy, patently useful woods must originally have been [...]ted by humans. Only large, ragged and often distant collections of trees are ever commonly [...]ught of as 'ancient' and therefore pristine.

It may seem precious to quibble about a wood's provenance. All treeland contributes to the vital [...]ocess of helping clean and cool the Earth's atmosphere; and a young wood, even a planted wood, will eventually become an old one. Yet a wood's origins do profoundly affect its character. There is still a widespread belief, dating perhaps back to the heyday of 'Improvement' in the eighteenth cen-tury, that the vast majority of domestic woods were planted by men, and that therefore the chief aim of woodland conservation is simply to plant new trees when the old ones are cut down. The unspoken assumption running through this, and much of woodland mythology, is that woods are replaceable and interchangeable; where there are trees there is a wood, and therefore – perhaps in some kind of ecological tautology – woodland plants and creatures: sheets of bluebells, elfin fungi, a nightingale in a thicket, and venerable oaks individual enough to have their own nicknames.

But experience increasingly suggests otherwise, and we are beginning to understand that woods with long ancestries have ecological characteristics that cannot be re-created in new plantations. They have greater ecological diversity, and a range of species (and often physical features) that do not occur in new woods. They also have a less tangible asset, in that they can give us a unique insight into that paradoxical intimacy between antiquity and new life that is one of the features of successful eco-systems.

But to understand how this can be, why ancient woods have these qualities, and how they have persisted into the present, we must first look at the history of our natural woodland, and how it was influenced by human activity.[4]

About 7,000 years ago, when the climate was warmer and human influence in the landscape was small, about two-thirds of the land surface of Britain was covered by woodland. Only wet, mountainous and unstable ground was free of trees for any length of time. The rest was the wildwood, the primeval wilderness. This 'deciduous summer forest', as ecologists have evocatively named it, was (and still would be, if humans disappeared) the natural, climax vegetation on these islands.

Yet this picture suggests something altogether too uniform, too *finished*. Both everyday experience and archaeological evidence suggest that the traditional image of a vast forest of mighty oaks, interspersed here and there with various inferior species, has more to do with sentimentality and those old mythologies than with fact. The best evidence of all is the immense variety of woodland types that have survived up to the present. They are not literal representations or intact relics of the wildwood of course, but its lineal descendants – part, so to speak, of the native woodland's family tree. Yet even in these reduced and modified forms we can catch an echo of the richness and diversity of our aboriginal forest cover.

There are beech woods in the Chiltern and Cotswold hills, ash woods on the Pennine limestones, pine woods in the Scottish Highlands (and oak woods as well, even as far north as Sutherland). There are a host of smaller, locally idiosyncratic types, too. Yew woods grow on the southern chalk, especially at Kingley Vale in Sussex (see p. 65), where in late winter their dark tiers – looking more black than green – are softened by the cottony-white seed-plumes of old man's beard. A scrubby expanse of holly and prostrate blackthorn (first mentioned in a tenth-century charter) hugs the shingle on Dungeness beach. At Staverton Park, in Suffolk (see p. 95), holly appears even more remarkably, in what amounts to a second-storey wood, rooted ten feet above the ground in the crucks of ancient oaks. There are serpentine alder woods in river valleys, birch woods on wet fenland and dry northern moors, and on the Taynish peninsula in western Argyll a wind-blown wood of juniper that is probably much as it was after the retreat of the glaciers 10,000 years ago.

But describing a wood in terms of its commonest or most persistent tree species rarely catches the local mixtures that give woods their character. In parts of western Scotland, for instance (as at Ariundle, see p. 135), the oak woods would be more properly described as sessile oak-elm-ash-hazel woods. At another extreme, in Bardney Forest in Lincolnshire, oak grows in intimate association with small-leaved lime. Many other East Anglian woods have complicated mixtures of lime with ash, maple and hazel. In one ancient Cambridgeshire coppice – Hayley Wood, which has a written history extending back to 1251 – Oliver Rackham has identified no fewer than six discrete tree communities (often known as 'stands'). They include aspen groves on waterlogged soils, an ash swamp, and 'the Triangle', a new or 'secondary' wood sprung up this century on an abandoned field next to the old wood.

The diversity of vegetation in ancient woods is matched (and in part caused) by natural variations in their physical structure. Woods can include slopes of strewn boulders, cliffs sheer enough to be quite free of trees, and clearings created by fire or the natural falling of large trees. Dead wood in one form or another must have been one of the major elements in the original wildwood, and is still a crucially important habitat for mosses, fungi, insects and insect-eating birds. (In some of the forests of Eastern Europe, believed to be the last remaining fragments of pristine temperate woodland, as much as 50 per cent of the timber is dead or dying.)

Water, too, can dramatically change the patterns of vegetation in a wood. A river creates a strip entirely free of cover, which may itself be fringed with open habitats like gravel beaches and sand banks. If the ground is poorly drained there may be fens, bogs and alder swamps, or damp meadows and grazed, grassy 'lawns', as in the New Forest (see p. 55). In some river estuaries in Cornwall there are tidal woods where, at the equinoxes, sea-water rises to lap the lichen-festooned oaks and primrose banks.

Frost, lightning strike, fire, even large animals (of which we are the most recent and influential example) all play a part in shaping the architecture of ancient woods. Deer, for example, not only beat out regular tracks, but can convert a temporary clearing to permanent pasture by browsing away regenerating shoots and seedlings. Left to its own devices (and provided with enough time and space), a wood will quite naturally develop a huge variety of internal habitats that range from sunlit glades, full of flowers, to the dark, aerial ponds that can develop in rot-hollows.

Woods evolve and develop in this way entirely of their own accord. Yet, from the moment that early humans began to establish settled communities and clear the trees to make room for cultivation, they began to use and therefore re-shape the forest. It was the source of fuel and building materials, and of a good deal of food. It provided browsing and shelter for their animals. While there was still a small population and an almost boundless expanse of woodland, these uses were compatible. If cattle grazed out young saplings and held back regeneration near the settlement, there would always be, just a little deeper into the forest, an abundance of trees of all shapes and sizes. But as the area of uncleared forest shrank, and the demands for usable wood became more intense and complicated, it was inevitable that some deliberate, albeit rough and ready, management principles would be worked out. There must have been two overriding needs. First, to prevent stock eating away the new tree growth as soon as it appeared. Second, to try and make this growth more regular and predictable.

It would be hard to say when the formal answer to this need – the enclosed coppice – evolved. But it cannot have escaped the attention of early wood-cutters that most of our native deciduous trees, when felled, regenerated naturally by sending up sheaves of straight new shoots from their stumps (usually known as 'stools', to distinguish them from dead stumps). It was then only a short step to deliberately cutting down trees, not just for their immediate timber yield, but for the plentiful supply of poles that would sprout from their stools; and from there to physically separating the woodland areas, where cattle were grazed, from the areas where coppice poles were grown.

The Bradfield Woods in Suffolk are the finest surviving example of a worked coppice, and parts have an almost unbroken history of coppicing going back to 1252, when the wood was owned by Bury St Edmunds Abbey. It looks much as it must have done 700 years ago (and is probably even managed in similar ways) and is an archetype of the coppice system that conserved so many elements of the wildwood.

The whole wood is surrounded by a huge boundary bank and ditch, a characteristic feature of ancient coppices, and at one time a vital measure to keep wandering cattle away from the growing shoots. The underwood was cut on a short rotation (about ten years), so that at any moment there were some compartments in the wood that were completely open and others at all stages of regrowth up to the dark and densely packed thickets, 30 foot tall or more, which were ready for the next cut. From notes in account books, it looks as if the composition of the underwood was much as it is today – a mixture of ash, hazel, maple, lime, birch, crab apple and willow.

Very little went to waste. The finer cuttings were bound up as faggots for firewood. Hazel was especially valuable, as it could be woven into wattle hurdles, for use in fencing and house-building. Selected specimens of species such as maple and ash, which supplied finer-quality timber for making domestic furniture or farm implements, might be cut on a longer rotation of up to 20 years. And scattered amongst these cut stumps, which would continue to provide poles indefinitely, would be a few 'standards', oaks especially, which would be allowed to grow on till they were 100 or so years old. Sometimes these trees originated from the selection and 'promotion' of promisingly straight and vigorous coppice shoots. Sometimes they grew from seedlings that became established in the open, sunny conditions of a recently cut compartment. Only very occasionally (and not until comparatively recently) were they deliberately planted.

These timber trees were often cut selectively, being taken out because a certain pattern of branching or a curve in the trunk fitted the needs of a particular piece of building work. In all kinds of ways the management of ancient coppices was integrated with the social life of the parish, from supplying its firewood and the raw material of its craftspeople to influencing the design of its buildings.

So too was the other major management system, 'wood-pasture' – though here it was the social organization that was complex and the management that was simple. Wood-pasture is a term used for a type of land on which the growing of trees is combined with the grazing of stock. The animals feed on the grass and flowering plants under the trees, browse the leaves and lower branches, and forage for acorns and beech-mast when these are available. There is, as a result, very little regeneration from seeds and suckers, and no new growth from existing trunks at any level within the reach of cattle.

The usual technique for reconciling tree-growth and grazing in pasture land was 'pollarding', the lopping of branches at a height of between six and ten feet above the ground. Like a coppice stool, the pollarded boll (provided it had enough light) grew new branches to replace those that had been cut. But the cutting cycle was usually less regular than in a coppice, as lopped branches were really only useful for firewood or rough timber. An important consideration regulating the length of the rotation was the need to keep the underlying grass well-lit and productive. (As a bonus, regular pollarding increased the life-span of the trees themselves by reducing their liability to wind damage and toppling.)

An old-style wood-pasture is a fair approximation of the popular image of ancient forest, and is the kind of woodland that can still be seen in places like Binswood (see p. 75), the New Forest (see p. 55) and Windsor Forest. It has a scatter of old trees, often sculptured in extraordinary shapes because of centuries of pollarding; dappled clearings; a rather sparse ground flora but a good deal of dead and fallen wood. This simplified picture belies the variety of administrative systems that regulated wood-pastures. Many were wooded commons, for instance, the wasteland of the manor over which rights to cut wood (though not to fell trees) were granted to the villagers. There were also enclosed parks, which usually contained cattle or red deer, and the Royal Forests. Here 'forest' was not a term describing woodland cover, but a legal definition, indicating that certain laws designed to conserve deer (and their habitats) operated over a particular area of land.

The details of ancient woodland management, fascinating though they are, give only circumstantial evidence about the origins of what we see in woods today. But they all point to the likelihood that both coppice and wood-pasture were modified relics of the original wildwood, carved out of the forest rather than specially planted; and that these in turn have survived into the modern era as what we call ancient woodland.

Considered by themselves, of course, neither wood-pasture nor coppice bears much resemblance to 'natural' woodland. But, added together, they represent most of the elements that must have comprised the primeval forest. What early woodland management did, in effect, was to split the two chief layers of a natural forest and conserve them under different management systems. The coppice preserved the continuity of the ground flora, the shrubs and young trees, and the bird and insect life that depended on them. Wood-pasture preserved ancient trees and dead wood and their associated communities.

But it would be complacent to assume that the use of woods automatically conserved them and their ecological variety. One of the less happy consequences of organized management is that woods become discrete and defined. In acquiring boundaries and names, they become separated from one another. This makes little difference to adaptable and mobile creatures like mammals, birds and flying insects. Yet it puts finicky woodland plants with poor colonizing abilities in a much more precarious position. If, for any reason, a species is wiped out in the wood, there is no easy way in which it can reappear by natural colonization. The source of new plants is simply too remote. The fortunes of species hanging on in isolated pockets of ancient woodland resemble those of species on islands, and the process of clearing the wildwood has been described as 'islanding' – a movement from a continental land-mass of forest, broken here and there by lake-like clearings, to a cluster of wooded islands in a sea of cultivation.

Yet historically it was the increasingly frugal use of these islands of ancient wood that helped to ensure their survival. By the Middle Ages woodland was too scarce and valuable a resource to squander. One of the most obstinate myths that has grown up this century – and which must bear some of the responsibility for our modern horror of cutting down trees – is that our native woodland was *destroyed* by being used: burned up in the furnaces of the iron masters and sunk with the great ships of successive battle fleets. It is an understandable myth, but not supported by historical evidence. Iron-smelters (or ship-builders, for that matter) who destroyed one of their basic resources would rapidly go out of business. They depended absolutely on a continuous, reliable, local source of wood. And the boon of the coppice is that it can meet this need more or less indefinitely. It is no coincidence that many of the largest surviving blocks of ancient coppices coincide with areas of one-time ironstone quarrying, or with other fuel-based industries. Wyre Forest, for instance, the Loch Lomond oak woods, Furness and the Lake District, the Wye Valley woods (see p. 105) and the Forest of Dean all produced charcoal for the local iron industry. The Chilterns exported beech and oak to London for fuel in medieval times, and later harvested pole beech for the chair-making factories in High Wycombe. Around Dartmoor in Devon, many oak woods (such as Black Tor Copse, see p. 25) were cut for their bark, which went to the tanning trade (extensive in this stock-rearing region). Even in Cranborne Chase (see Ashcombe Bottom Woods, p. 45) much of the hazel coppicing was done for local hurdle-making businesses.

All our native trees will regenerate from stump, seed or suckers, provided they have sufficient light and are protected from browsing animals. Even a wood that is clear-felled will have a rough coppice crop harvestable in ten years' time, and useful timber again in 50. It was not employment that destroyed most of our ancient woods, but redundancy. If a wood was felled at a time when corn or conifers, sheep or speculative building were more profitable than native timber or underwood, then the cutting of the trees was often followed by the clearing of the ground. The use of the land itself changed.

There is no doubt either that low-intensity, sustainable management can increase the ecological diversity in a woodland. This is partly because it can change the balance between different species and communities, but chiefly because it can compress and accelerate the kind of habitat changes that occur naturally in a wood. A recently cut coppice plot is, ecologically, much like a clearing created by a large wind-blown tree. Artificial rides and tracks extend the kind of open, linear habitat that occurs along the edges of streams and deer-lawns. Steep banks by the sides of tracks, or along the boundary of a wood, can provide fair imitations of well-drained, small-scale cliffs.

Even individual trees incorporate both natural and human 'weathering', almost without distinction. Everything that happens to them – lopping and nibbling; attacks by lightning, wind, woodpeckers, graffitists; sapwood-fattening spring rains and shading by planted competitors – is incorporated, in-grained literally, in their structure. As they mature, trees become increasingly complex elaborations, not just in their ever more intricate branching, but in their accumulating array of knots, burrs, scars, bark-reticulations, holes, layers of lichen and moss. It is impossible to measure the area of a tree's surface decisively. It is what is known as a 'fractal' quantity – one that increases indefinitely, the more closely you examine it. What all this amounts to is that ancient woodland (especially 'primary' woodland, which is assumed to have a continuous history as woodland back to prehistoric times) is irreplaceable. Its structure, vegetation and many of its less mobile species, are dependent on that continuity, and promises by developers to 'restore' a wood that has been destroyed by, say, quarrying, should be viewed with scepticism. They may re-establish the trees but they cannot re-create the wood.

But there is a paradox at the heart of ancient woodland history . If management has helped to save and conserve the wildwood, it has also been responsible for destroying it – or at least, almost by definition, emasculating its essential wildness. There are good social and even ecological arguments for continuing traditional management practices in places where natural resources are scarce. They guarantee continuity of conditions for those animals and plants already present, preserve historic features, involve human communities and generate income. But what they also do, as we have seen, is irrevocably separate two kinds of woodland – the underwood run as coppice, and the old trees as wood-pasture – which in a state of nature are intimately mixed. As a result, there is almost nowhere in Britain where we can see what a *complete* mature native wood is like, and we have only shadowy ideas, from examples in other parts of the world, of how such a wood might change and regenerate over the centuries. We are not sure, for example, how a large group of dead trees would finally collapse in the kind of conditions that prevail in Britain. Would they go over separately, in order of age or height, or according to the whims of the weather? Might they prop each other up until they were all sufficiently rotten to fall together? Would natural openings in a previously dark wood fill with flowers in the spring, like a coppice compartment after cutting, or with rank undergrowth? Would seedling trees take root in the stumps and dead wood, as they do in warmer parts of the Continent?

The best opportunity for regenerating a genuine wilderness or wildwood in Britain would be in the Scottish Highlands, where the human population is thinly scattered and soils are infertile. In the past, ironically, it was just such conditions that made the Highlands easy prey for all kinds of exploitation. It is one of the rules of the 'progress' game: vast reserves of empty land attract people eager to appropriate and misuse them. In the Highlands the process has been especially damaging because the usurpers have so often been outsiders with grandiose ambitions. In the seventeenth and eighteenth centuries the native pine forests – one of the last extensive remnants of the north European wildwood – were looted for the English navy. In the nineteenth century it was the native inhabitants' turn, and the Highlands suffered the terrible injury of the Clearances. In Sutherland alone one-third of the population was evicted (by burning down their homes in most cases) to make way for sheep-ranching and game-shooting. Sheep-grazing and moorland-burning (for grouse) ensured in their turn that the pine forests would be unlikely to regenerate. In the twentieth century much of this open country has been put under another foreign régime – huge shrouds of alien conifer species that for much of the 1980s gave tax relief to absentee entrepreneurs and speculators. Now power and ownership have drifted even further offshore, and walkers and birdwatchers just up for the day from the Lowlands can find themselves fenced out of the moor by the multinational grouse-shooting and deer-stalking syndicates.

But recently the Highland bubble has begun deflating, and so much wild country is changing hands (or at least changing status) that it has begun to look as if Scotland might at last be able to possess a sanctuary free of commercial pressures, where native wildlife could flourish and humans meditate upon their own species' follies: a kind of wilderness, in fact. I use the word with hesitation, since in these days of global tourism and ubiquitous pollution it is, to say the least, a somewhat relative term. It is also a contentious one, which tends to undervalue the presence of humans in remote areas and overvalue their contributions elsewhere. Yet in a country where even nature conservationists often seem bent on taming every last acre, I think that the time may be ripe for bringing the idea – and in most of Britain the term 'wilderness' means some kind of naturally evolving wildwood – back into currency.

It is not an idea that meets with unqualified approval up on the estates. In parts of Scotland the sanctity of traditional ownership is still regarded as an inviolable principle. The 'stewardship' and 'caretaking' of nature are spoken of as if they were universally and self-evidently desirable aims. Even in these great spaces we seem unwilling to let go of the reins.

This is not to deny that there are real practical obstacles to establishing wilderness woodland in Scotland. The most intractable obstacle is probably the red deer, which in most remnants of the native pine woods hinder regeneration by browsing away pine seedlings. Deer numbers have been climbing steadily ever since the animals became a valuable resource for owners of stalking rights, and in many areas of the Highlands they are now at levels that the land cannot easily support. One obvious and frequently canvassed solution is regular, heavy culls. The distinguished Scots ecologist, Adam Watson, believes that the native pine woods cannot even survive without a great initial reduction in deer numbers, followed by a regular programme of culling. His cautionary picture of what is likely to happen if there is no great cull is not a pleasant one:

Nature will produce her own solution with mass starvation. I have seen hundreds of dead adults and calves in woods and treeless glens after a snowy winter. They were emaciated, with their ribs sticking out, and in summer their decomposing carcasses made the glens stink.[5]

It is a discomfiting picture, but it is at least a 'natural' one., This is the way that wild animals do die in times of hardship, and it might do us no harm to witness, on our doorsteps, the consequences of our past short-sighted plunderings of nature.

The red-deer problem is just one example of the quandary that faces anyone who wishes to establish wooded wildernesses in Britain. The 'naturalness' of almost all wild areas has been compromised by past human action. Exotic mammals have been introduced without any natural predators to control them. Exotic trees compete with the natives. Diseases, pollution and human recreational pressure can, if unchecked, threaten precisely that ecological richness that a wilderness was established to conserve. Does one accept this risk, or compromise the wilderness from the outset by managing it?

These days of course no definition of wilderness could reasonably include an absolute absence of human interference. In his seminal study *Wilderness and the American Mind*, Roderick Nash makes the point that wilderness is more a state of mind, 'a perceived rather than an actual condition of the environment'.[6] To one child to whom he spoke, wilderness was 'the dark space under my bed'. To Wordsworth, who wrote some of the earliest and stoutest defences of the concept (including a damning comparison between plantations and naturally sprung woods in the Lake District), it was a condition of the spirit as much as of the land.[7]

Yet there must be a base-line in this debate. It is possible to argue about how much a place is compromised by its past history, and by current, accidental human influence. But the idea of wilderness surely becomes meaningless unless it implies a state of affairs in which, at a specific point in time, over some quite large tracts of land, *deliberate* human control has been largely removed. A wilderness in meaningful only if it is a place of contrast, where the rules and values of normal land-management are challenged, and where we can learn something about humility and the resilience of un-managed nature. And, as part of this, there must be the possibility of things turning out quite differently from expectations, which means accepting possible degradation as well as creative surprise.

It is possible to imagine how this kind of definition might be translated into an action plan for the remaining fragments of the Great Wood of Caledon (as the remains of the native pine forests are collectively known; for example, Glen Strathfarrar, see p. 165). Initial fencing and the removal of roads, for instance, would lie on the acceptable side of the base-line. But regular programmes of 'calf and hind management', pest control and the removal of aggressive alien tree species would, I think, be beyond the pale. They are entirely acceptable in nature reserves with clearly defined goals, but not in a wilderness.

The question is whether we can afford to take such risks anywhere in Britain, given our limited space and natural resources. Many would argue that the overriding aim of conservation is the saving of species and specific eco-systems, regardless of how much un-natural management this may entail.

But there is an alternative point of view. In *The End of Nature*, the American writer Bill McKibben argues that nature traditionally held crucial, non-material values for us, but that with the onset of man-made global warming and the spread of genetic engineering:

> our sense of nature as eternal and separate will be washed away . . . . Before any redwoods had been cloned or genetically improved, one could understand clearly what the fight against such tinkering

was about. It was about the idea that a redwood was somehow sacred, that its fundamental identity should remain beyond our control. But once that barrier has been broken, what is the fight about then? It's not like opposing nuclear reactors or toxic waste dumps, each one of which poses new risks to new areas. This damage is to an idea, the idea of nature, and all that descends from it.[8]

McKibben attacks, too, the idea of our becoming stewards of a managed world, 'custodians' of life. 'For that job security we will trade the mystery of the natural world, the pungent mystery of our own lives and a world bursting with exuberant creation?'

McKibben is writing about a global crisis. But his defence of an independent natural creation seems to me to apply just as cogently to our home woodlands. Unlike America, the idea of unmanaged wild places is not viewed sympathetically in Britain. There are good reasons for this. Our landscapes are mostly small-scale and have been fashioned by centuries of human work. There is, too, a powerful argument that the conservation of species (*not* of wildness) is best achieved by maintaining traditional patterns of management. Yet the case is not always put as calmly and pragmatically as this. At times it seems to be purely nostalgic, based on a naive belief that there really was a rural Golden Age when humans and nature were in harmony; at other times it is tinged with a more ugly neo-colonialism, a conviction that wildernesses are all right for the Third World, but not for us civilized folk.

The fear of 'letting nature go' reaches very deep in Britain. When, a few years ago, the idea of 'set-aside' was first mooted as a solution to the problems of surplus arable land, all kinds of bizarre fears began to surface about the British countryside being 'swallowed up'. All that was being proposed was a slightly modified revival of the old fallow system, and some modest experiments in farm woodland. But the end result, the soothsayers warned, would be a nightmarish waste of swamp and scrub, thick with rats, disease and the remains of derelict farm machinery.

What actually happens if a field is permanently abandoned (as countless have been throughout history) is a matter of simple ecology and physical record. After a few years of rank weeds, followed by brambles and scrub, young forest trees begin to appear, usually oak, ash or birch, depending on the underlying soil. Within a decade or two the field is unmistakably a young wood, with a more natural aspect (and probably more chances of survival) than any group of planted trees.

What precisely happens when an ancient wood is abandoned is, as we have seen, less certain. There are only a handful of deliberate 'non-intervention woodlands' in Britain, and they are all comparatively small-scale experiments, which have been under way for a few decades at most. The evidence from ancient woods that have been left more or less un-managed by default is mixed. They lose some species and gain others. Over the short term, at least, they often become less densely covered by trees. All that one can say with reasonable certainty is that a wood which is abandoned, which is allowed to become feral, will (unless grazing pressure on it is exceptionally heavy) continue to be a wood of some kind. Yet the myths about its likely future are almost identical to those that persist about abandoned farmland. By some mysterious process of degeneration it will simply cease to exist. In its place will be that loathed, amorphous habitat, *scrub*. In this fearful version of ecological history it is scrub, not wildwood, that is the climax vegetation in Britain. The distinguished landscape architect, Nan Fairbrother, warned in revealing phrases that: 'Incipient scrub always lurks, however, only temporarily suppressed: it is the state of original sin in our landscape.'[9] I sometimes wonder how those that perpetuate this myth imagine that Britain cloaked itself with woodland 10,000 years ago, long before the invention of foresters.

The evidence from real wildwoods is that they are more resilient than most of us (ecologists included) assume. One of the classic cases in Britain is Lady Park Wood (see p. 105), an ancient and very mixed coppice in the Wye Valley, which has been conserved as an unmanaged reserve for nearly 50 years. Up to the 1970s it was presumed to be proceeding towards the beech high forest that ecologists believed was the stable climax vegetation on limestone. Then came elm disease, followed smartly by the drought of 1976, which killed off many of the beeches. Gales and hard winters followed, including one severe ice-storm, which welded coppiced small-leaved lime branches to the ground – where they promptly took root. Lady Park is now dominated by survivors, especially ash and bizarre lime trees, and has become an exciting, vital, wood, which has transcended the future that was conventionally predicted for it.

Yet we do not really believe such things unless we see them, and even then we do not always credit them. After the '87 hurricane I watched bulldozers in Sussex clearing acres of healthy and naturally sprung seedling trees in order to plant nursery-raised specimens. The spontaneous seedlings were not recognized or accepted as real trees; being unplanted, they were 'weed trees'. Whenever woods do not behave in the ways that we expect, we fall back on this kind of language, borrowed from the vocabulary of human domination. Foresters talk repeatedly about *dereliction, overmaturity, scrub* – as if trees, the most successful and durable of all plants, were incapable of living correctly without supervision.

I suspect that the myth of the doomed future of the unmanaged wood is, at root, a fear of the fecundity of nature; of what John Fowles called 'green chaos'; of the wildwood itself. It is a fear that has run through our culture parallel to our love of woods as places (and symbols) of sanctuary and creativity. In this respect we may not have advanced much since the so-called 'Age of Improvement'. In 1712 John Morton wrote that 'In a country full of civilized inhabitants timber could not be suffered to grow. It must give way to fields and pastures, which are of more immediate use and concern to life.' It was an attitude that Europeans took with them to the New World. When William Bradford stepped off the *Mayflower*, he described the North American landscape as a 'hideous and desolate wilderness' and promptly set about bringing it to heel, just as his ancestors had done so successfully throughout Europe. Within two centuries the descendants of the original settlers had obliterated seven-eighths of the North American continent's natural woodland (and whole species like the passenger pigeon) in pogroms of an arrogance and violence that rival those in modern Amazonia.[10]

Yet few of us are in a position to be self-righteous about this. My own wood in the Chilterns is sporadically managed, and every winter we go in and knock over a few trees. It is all quite proper, we assure ourselves, just thinning and coppicing, making space for new growth. The cut wood is all used, and the local wildlife patently benefits. But there is more to it than is apparent in these very reasonable justifications. There is an exhilaration that can grip you when you start working in a wood – the satisfaction of a clean fell, the tang and feel of fresh-cut wood. I have seen our helpers, friends of the wildwood to a soul, sawing and stripping felled trees with the single-mindedness of jungle ants reducing a dead animal to a skeleton.

Woods – ancient woods especially – are a *challenge*. Their longevity and sheer stubborn rootedness spark off all manner of competitive and resentful responses. To the Puritans, the wildwood was a barbarous offence, something not yet in a state of grace. To a modern forester, it is undisciplined and unproductive, an insult to his skills. To a farmer, it is a waste of tillable land. Even to ecologists it is a goad, challenging them to crack its codes. I am apt to defend the opening out of my Chiltern patch as helping to speed the wood's 'natural' development – enlightening it, so to speak. But I fear that I may share more of those old proprietorial attitudes than I would care to admit: all of us – apostles of the New Woodmanship, Malaysian logging barons, inheritors of ancestrally wooded Cotswold estates – share the same self-interested belief: that our plans for the wildwood are better than its own.

But perhaps this is being too severe. It is also that our plans suit *us* better. To suggest that it would be fascinating and culturally rewarding to allow some areas of our native wildwood to 'go back to nature' is not to say that we should stay out of the remainder. We have a long history of living and working with woods, and as much right to continue this as any other forest creature. It is the *style* of our involvement, and the assumptions behind it, that are the real issues in most ancient woodland, rather than a stark polarization between comprehensive management and total non-intervention.

Working through these apparent contradictions is unavoidable when you have notional charge of a wood. When I first bought my 7-hectare patch in the Chilterns in the early 1980s, I was full of idealistic plans for being the Good Woodward. I would create a sanctuary, nurture the plants and birds, grow timber without compromising these aims, and return the Greenwood to the People. They were worthy enough aims, but for a while I half-convinced myself that none of them could be achieved without putting the wood under something close to intensive care.

A decade on, I am wiser about my own presumptuousness and about the natural world's ability to look after itself. The wood, I believe, has flourished under our care, and some of those grand aims have been realized. But its current state is only one of a number of lively, viable futures that it might have had, and the one that happens to please those of us who use it. Meanwhile the wood has continued to have a life of its own, often in quite unexpected ways and places. The badger tracks are as wide as some of our footpaths. Naturally regenerated trees have swamped most of our plantings. And some of the least-managed parts of the wood – especially those that were hardest hit by the great gales of '87 and '90 – have evolved into natural glades of the kind that we often spent months of arduous work trying to create.

Weather and seasonal change cloak the whole place with a continually changing and unpredictable patina of wildness. In winter the whole interior of the wood can be transformed. Rain storms leach out banks of flints and miniature flood plains. Drifts of snow or leaves can obliterate existing paths and leave new trails of open ground between the trees. I walk about as if I have never been in the place before, seeing the other sides of familiar trees. Snow and hoar-frost can weld the outer branches of hollies to the ground, forming a kind of protective skirt under which pheasants and woodcock sometimes shelter.

The natural pulse of the wood can show through in the smallest corners. There is one small, deeply shaded patch at the top of the wood where the soil is very thin and acid, and where the vegetation seems to have reached a tacit agreement to share out the meagre resources. There is barely a hint of the struggle for survival. The wood anemones keep to their place, close to the clay, and the melick grass to its place, up on the wood-bank. The bluebells never grow more than six inches high, or the wisps of honeysuckle more than a foot. The only sign of change I have seen here was when the leaves of a bluebell, normally content with spiking dead beech leaves and carrying them a few inches into the air like starched brown ruffs, broke through the rotting shell of a fallen branch. In this still centre of the wood this was a major event.

Our own attempts at landscaping have also been accommodated and softened by natural forces. The most extensive, the main track that I mentioned previously, seemed to be naturalized almost as soon as it was made, despite being created by a huge agricultural excavating machine armed with a prehensile scoop. The operator seemed able to use this like a precision instrument. He would excavate scoopfuls of clay and tuck them underneath the machine in order to raise his working platform. He added a subtle S-shape to the crudely direct route we had marked out up the hill. I became fascinated by the machine, and in my excited diary entries for those days it had already been gathered into the company of honorary forest creatures:

> When he is dragging soil back to level out the track, he pulls it back and under himself, as badgers do when dragging bedding out of their setts . . . all the while the machine being followed by small bands of robins, like tiny gulls after a woodland plough.

The end product was remarkable, a sinuous, unsurfaced road lit up by the late afternoon sun throughout much of the winter. Even after one week its hard edges were beginning to mellow. The first frosts and rains brought down drifts of debris over its bare surfaces – bluebell bulbs, moss, flints. Within a month there was a green film of algae over the chalk, and the first weed seedlings. One year later the bank was already a recognizably Chiltern slope, tufted with musk mallow, old man's beard and wild basil.

But what struck me was that this generous flooding back of life had happened as much in spite of our landscaping as because of it. We had created a framework, opened up the soil, no doubt inadvertently moved seeds about. But what filled the landscape, gave it life, texture and surprise, was nature's contribution.

I noticed this same sense of locality, and a similar blurring of the effects of human management and natural maturing, when I developed a taste for exploring woods, back in the mid-1970s. It was a kind of wood-crawling, I suppose, sometimes involving nothing more than a short foray from the road when I passed a likely-looking copse, sometimes long, cross-country journeys to hunt down a particularly choice inhabitant of a highly local group of woods.

In Kentish coppices I listened to whole ensembles of nightingales, and in a Dorset deer park I walked round Billy Wilkins, an 800-year-old oak pollard, named after a local family and so vast that its different aspects are like quite different landscapes. I saw ancient ash trees rooted directly in the cracked limestone rocks at Colt Park Wood in Yorkshire (see p. 125) and so starved of nutrients that they were growing like bonsai; and 1,000-year-old ash stools in some East Anglian woods that were 15 feet or more across, hollow in their centres but still vigorously alive.

There were other still-extant coppice remnants, including some 1,000-year-old small-leaved limes in the Pennines – only their stools had the look of pollards, because the soil level round them had, over the centuries, dropped by many feet. Limes proved to be a perennially fascinating tribe, and I sought them out in all kinds of woods, as billowing maiden trees in remote woods on the South Downs, and as coppice in big commercial woods in Lincolnshire.

Scarce trees like this are often regarded as 'ancient woodland indicator species', poor colonizers that grow successfully only on sites which have been continuously wooded. They include widespread species like wood anemone, sweet woodruff and lily-of-the-valley, but also local specialities like the true oxlip, which is more or less confined to coppice woodland on the boulder clays of Essex, Suffolk and Cambridgeshire. They seem as tangy and distinctive as local dialects, many of these constellations – wild daffodils colouring the horizon gold in the leafless woods of

Herefordshire in March; and, at the other end of the year, meadow saffron's satin-sheened petals twisting amongst the fungi in the Mendips. But they can also form common bonds of continuity between very different ancient woods. I have seen herb Paris, with its extraordinary crown-like flower, growing under dappled hazel coppice in Hampshire and deep in a crevice on a Yorkshire limestone pavement, where it may have been confined for centuries. In my own wood, one ancient woodland plant has prospered above all others – along the edge of our new track. It is wood vetch, a straggling climber that carries bunches of sweet-pea-scented, lilac-striped flowers up into the bushes and the lower branches of trees. In other parts of England its natural habitat is rocky cliffs and river banks in woods, but after we had cut our track it spread from a small, single plant to drape the whole length of the new, sun-lit embankment.

Ancient woodland plants are by no means confined to untouched or wilderness woods. What their continued presence indicates is that the human activity in a wood is tolerable to it, maybe even adding to its diversity, in the way that beavers or badgers do. Humans will never be wildwood indicator species themselves, but, modestly feral, they can be acceptable denizens.

One of the fascinations of Gareth Lovett Jones's work is the way it explores this common ground between natural change and the aftermath of human activity. His pictures are full of spaces – the areas and moments where change occurs. Some are where trees have been cleared: deliberately, as at Binswood (see p. 75), for the sake of grazing animals; or naturally, as in Pinnick Wood (see p. 55). It is hard to tell the difference. Many are not obviously in woods at all, but show stone, water, even sunlight, turning into the stuff of trees. In the humid depths of Black Tor Wood (see p. 25) and Ty-Newydd (see p. 115), boulders are draped by pioneering mosses and ferns. In the open areas of Eilean Suibhainn (see p. 155) and Colt Park (see p. 125), they are followed by young trees, rooted sometimes in dead wood, sometimes in the rock itself. In the Ariundle Oakwood (see p. 135), branches fill the spaces between trees, and seem to *wind* into them. Time – briefly frozen, but implicitly flowing up to and beyond the picture – is a crucial ingredient. One portrait of a gully in Glen Strathfarrar (see p. 165) is described, typically, as 'taken in a pause between the lightest of snowfalls'. In a remarkable sequence of studies in the Chilterns (Lovett Jones's home country – see pp. 88–94), beech trees move in the rain and fog, catch a moment of light breaking through from a nearby clear-fell, and, striping each other with trunk shadows at sunset, are inextricably linked as a community, even in their use of the basic resource of sunlight.

One picture, a bird's-eye view of a group of bleached, fallen elms at Ashcombe Bottom (see p. 45) is uncannily reminiscent of the famous hand-drawn map of fallen and standing trees in a plot of the ancient and near-pristine forest of Boubinsky Prales, in southern Czechoslovakia. First surveyed and mapped out by Joseph John in 1847, and updated each year since 1954, this map shows in a series of 'snapshots' the details of the infinitely slow decay and regeneration of the wildwood. There is, I think, a close analogy between records like this and the growth-rings of trees, those year-long time-exposures which record – just as Gareth Lovett Jones's photographs do – the detailed, day-by-day experience of the woods in which they are developing.

In Search of the Wildwood

– BY GARETH LOVETT JONES –

The photographs in this book record some of the features of fifteen widely spaced patches of ancient woodland. I must admit to having been guided to some extent in my angle of view by what I learned about each site, but essentially this is an intuitive, not a scientific, record, and my main aim was to do some justice to the visual qualities that give each wood its character. In selecting the woods my criteria were, firstly, that they should be located over a wide range of British topography and, secondly, that the choice should convey, as far as is possible in a short book, an impression of the visual diversity – the forms of beauty, if you prefer – to be found in surviving ancient woodlands.

The pictures therefore cover a range of ancient woodland types, from wood-pastures such as those of Binswood and Pinnick Wood in the New Forest, through stands of ancient trees such as the yews at Kingley Vale in Sussex, or high-level primary oak forest such as Black Tor Copse on Dartmoor, to ancient ash wood, as at Colt Park in the Pennines and Rassal, Wester Ross. Luckily, and on reflection unsurprisingly, in this subject there is a direct correlation between scientific and visual diversity. To put it another way, with few exceptions, if a woodland is known to be of outstanding scientific interest or a rare example of a type of plant community, then it is bound to be very much its own place, and rich ground for photographic exploration.

I had photographed experimentally in four of the woods before work on the book began. I had few clear ideas about the rest, only a vague hope that in visiting and taking pictures of them I might perhaps be able to get one step closer to some picture of that vast ghost of a vanished primeval forest which – in the right places, and the right lights – still seems to haunt the British landscape. In his seminal work *Trees and Woodland in the British Landscape*, Dr Rackham acknowledges that very little is known for certain about the appearance of the 'wildwood'. But he does go on to speculate that it may have 'contained many very old and large trees, with little underwood, and large quantities of rotten trunks lying on the ground or leaning against trees', and that 'in many places the big trees were patchy or scattered, allowing light to penetrate the underwood'.

This composite picture remained with me throughout the months of exploration, and every so often – by chance, it must be said, more often than by intention – I did find myself confronting a scene that seemed in one way or another to marry up with it. This was not the wildwood itself, of course – how could it be? But it was always a plant community which, for all its standing in the present, possessed sufficient biological links with a distant past, and was sufficiently little influenced by the activities of forestry, to make a passing imitation of the condition of the wildwood – or at least (we must be precise!) of what the wildwood *might* have been.

This notion may well have been fanciful. Even so, I think it was useful to travel with preconceptions. I went to the woods expecting anything but the kinds of anodyne prettiness to be found in the country magazines and certain kinds of guide book, and made it a part of my business to eliminate all signs of the present-day activities of that 'influential' mammal, man. (A token presence was allowed in the final selection of pictures.) *Beautiful* the primeval forest could not have failed to be: but, as often as not, this beauty must have sprung from processes of decay, malformation, seeming-chaos and death, of parasitism and epiphytism, of the clusterings and overlappings of species all in competition for the same acreage or the same half-inch of territory. It is the beauty of what one disgracefully unreconstructed farmer known to me liked to call the 'rubbish' woodland: the beauty of scrubby, weather-twisted, fern-hung little oaks, or of distended, stag-headed or half-rotted ancient trees. I went to the woodlands hoping and expecting to find such abstracts, and I think that to some extent I may have selected them out from the more conventionally acceptable manifestations of sylvan beauty that were also, in some places, to be found there.

In the notes that follow, I have written about the woods from a number of perspectives, each of which is, I hope, connected with the others. Since my expertise as a botanist stretches no further than being able to distinguish a hornbeam from a beech (on good days; and only then if I am warned well in advance that the wood in question has hornbeams in it), and since there is now an abundance of specialized texts on the subject of ancient woodlands, it is as dangerous as it would be redundant to make detailed observations along these lines. Such botanical information as is included is drawn largely from notes supplied by managing bodies such as English Nature, where they were available. I have also deliberately avoided identifying any bryophytes, fascinating as they are, and have treated them only in the most general terms.

It seemed potentially much more useful to give the reader an uncluttered account of my own experience of the woods as landscapes – though perhaps I should say as *places* in which I walked. These are the ancient woodlands as they were seen in the most pragmatic terms by one individual, during a number of visits. Since that individual also happens at times to be a photographer, and since photographers seem rarely to write about the psychological or experiential aspects of their work except in private diaries, I have also included some account of my own role within each scene, and I have tried to be honest about it. There is, I think, a certain comedy in photography.

The Wildwood began with the notion of exploration, of a 'venture to the interior' of the British landscape through which a picture could be constructed of its most verifiably ancient natural elements. For this reason – if you like, as a part of that exploratory process – I have also incorporated some information on this history of the woods, where it was available.

Storm light: oak trees rooted amongst the clitter.

Black Tor Copse

LATE APRIL

A small woodland (four hectares approx.) of stunted pendunculate oak on Duchy of Cornwall
land in the valley of the West Okement River, some five miles south-west of Okehampton,
Devon, managed and monitored by English Nature (formerly the Nature Conservancy Council).
A tiny fragment of the ancient Dartmoor Forest, and probably primary woodland, the wood
stands on a valley slope covered in granite boulders ('clitter') beneath Black Tor. Of outstanding
interest both as an example of oak forest surviving under extreme conditions and for its
population of lichens and bryophytes, it is listed as a Grade One site in English Nature's *Nature
Conservation Review.*

The wood comes up like an island, very slowly indeed, across a gently sloping sea of black bogs and
bleached-blond grasses awash with this morning's rainwater. Even in April this is a dark island, a
slightly sinister clustering together of unquantifiable vegetation, holding its ground against the bar-
ren sweep of the moorland valley. Only when the cloud thins suddenly is this shadow-zone subtly
differentiated into lines of brown-black trunks and branches, their pale fawn branchlets pinkened at
their tips by the finest powdering of new-formed buds. In lowland Devon three miles to the west it
is a lush green April, rolling on into May. But up here on Dartmoor it is more like March; although I
do see one mare strolling easy with her foal, a little fine-limbed creature lost amongst great leaning
boulders on the hillside beyond the river. A gale is whipping down the valley, adding to my difficul-
ties: it is not until I get a lot closer to the wood that I see that there is a path of sorts, which I have
missed, down by the river's edge. But my own route has brought me closer to the true spirit of the
place: no doubt I would not have felt quite the same about it, had I not waded my cheery way up to
it across the sucking peat.

Rapidly over the last 200 yards or so the wood grows beautiful, fantastically detailed and eerie,
all in equal measure. The clitter itself begins to be pinkened by colonies of lungworts and lichens.
There are no fences or boundaries of any kind here and immediately I pass the first trees, teetering
clumsily on the smaller boulders, I see the same growths everywhere on the southern faces of the
rocks. Superficially the place resembles certain old industrial workings: it is like an abandoned rock-
dump, left to grow green under a score of different mosses and such trees as might succeed in
inhabiting it. But these rocks have been left here by forces somewhat larger than any to be tapped

by early industrialists. The oaks grow twisted amongst the boulders – was there ever such a tangling wood as this? – as if some malicious child-giant had been tweaking their trunks and branches, and bending them round like grasses until they were close to snapping. These wild and crazy twistings are in many places softly plumped out by brilliant green mosses, or serrated by stands of tiny ferns. Thus, a few of the more obvious effects on oak woodland of the altitude and the harsh Dartmoor climate, with its average rainfall of 84 inches per year, its low clouds and mists, and its winds.

At the southern end of the wood, almost before it has begun in earnest, a place has been made by God amongst the rocks for photographers to rest. It is one of those hermit's retreats such as are to be found only in the remotest upland wilderness, and comes equipped with a bone-dry recess for cameras and a variably dry stone seat for the artisan himself. I make this my base – rain is fast approaching – and go off to explore. Gathered between the boulders is a gorgeous humus, the colour of dark chocolate, and in the wettest places it gives succulently beneath my boots. The woodland is only a few hundred yards deep from east to west, and through the trees to the east I can see the Black Tor that gives the copse its name, with its grey strata stacked slightly skewed and overlapping. Beyond it lie the army ranges: I have long since realized that such close juxtapositions of rare wilderness and military preserve are commonplace. Just then, I see at my feet an ancient, burst-open shell, turning to rust in a clump of grass.

My first impressions notwithstanding, it takes me no more than a few minutes to enter into an understanding with the place. For all its bizarre conformations, for all that the trees themselves offer but the slightest protection against lashing rain storms, this seems a gentle place. There is no threat here (there is no threat now in any British woodland, unless it be from the activities of other human beings), and as the sky darkens once again I find myself half-hypnotized by the shushing sounds coming up from the river, as it negotiates its narrow bed of lichen-dappled boulders 200 yards away.

The wood is dominated almost to the point of exclusion of other trees by the pendunculate oak – Robur the Conqueror, indeed – but at present it is anybody's guess whether the rest of the Forest of Dartmoor was similarly constituted. It seems likely that most of the moor was under forest of some kind until about 2,500 BC; but it was cleared early, by the end of the Bronze Age, and such areas of woodland as were left standing then were subsequently cut back as fuel for tin smelting and other industries. Only two other, equally tiny fragments of this great forest survive today: they are the more famous (and thus more unfortunate) Wistman's Wood, in the centre of the moor, and Piles Copse, in the south. Evidently there were few serious attempts at wood-management on the moor, although the naming of two of these survivors as copses (that is, coppiced woods) may give a clue as to the reason for their continuing presence. Certainly some of the Dartmoor oaks were managed for their bark. Such considerations aside, however, there can be little question that in all these sites one is standing in woodland whose lineage from the wildwood remains unbroken, if not unmodified.

Of the three fragments, Wistman's Wood, being closest to a road, has been the most studied and recorded. Two facts in particular emerge from the evidence available. The first is that over the centuries – like Birnam Wood, but in this case without the assistance of soldiery – the wood has moved about in the landscape, changing both in size and outline. The second is that it has altered internally. As late as 1912, it was impossible to penetrate the wood except on one's knees: the trees grew very close to the ground, and many of their branches rested directly on the boulders. They were so thickly covered by epiphytes that, by one account, they seemed swollen to several times their true size, and long curtains of one moss, *Antitrichia curtipendula*, hung down from them, making it impossible to see through.

Today, Wistman's Wood is a very different place. During the past 80 years the trees have not only grown but lifted up from their boulder floor, so that livestock, and people, could penetrate the wood easily for the first time. The resulting increase in grazing has greatly reduced the number of ground flora and epiphytes, but fencing experiments in the wood have shown that keeping the sheep and cattle out does not necessarily encourage the oak saplings to regenerate, at least in the short term. What it has encouraged here is a vigorous regrowth of brambles. Outside the fences, however, it does seem likely that a 'moderate' amount of grazing can be of benefit to the bryophyte population, since competition from other species is thereby reduced.

The issue of grazing levels is crucial to the future of many of the most important ancient woods, especially in the uplands. But more research needs to be done, and effectively collated, before any definite management patterns can be established. As the English Nature leaflet for Wistman's Wood points out, 'it might be quite wrong to attempt corrective measures' of any kind. Even so, there is a clear case for fencing off individual saplings at Black Tor Copse, especially at the wood's margins. It is apparent that the wood is trying to spread outwards on its eastern side, where saplings can grow a few inches or even a few feet high under the protection of the clitter. But once they emerge into the open air they are browsed, and while a few of them are five or six feet high, the vast majority subsist as tiny low-lying shrubs creeping along the ground like heathers. Their fresh branchlets are grazed back to a level with much the same insensitivity to the principles of pruning as is displayed by the rotary flail as it hacks back branches in a hedge. There can surely be no harm in encouraging these woods – miniscule pockets of a once vast forest as they most probably are – to claw back a little of the emptiness that surrounds them.

Boulders at the wood's northern margin: reverse views. . .

. . . The growth of liverworts is especially profuse in the Dartmoor woods.

Looking across the centre of the wood towards the West Okement River.

Scene of a death: the pecked-clean skeleton lies 20 yards away.

The oak's branches here can resemble hanging gardens of fern.

November, before rain: two of the largest surviving beeches.

Lewesdon Hill

November and April

Eleven hectares of mixed ancient woodland, dominated in part by beech, covering the highest hill
in Dorset and surrounding the site of an Iron Age fort. Situated one mile to the south of
Broadwindsor, and owned and managed by the National Trust.

I climb the hill at four o'clock one sombre November afternoon, free from cameras, having been here
just once before, fifteen and a half years earlier. I remember the place vividly as supporting a collection of magnificent beeches, grown out into the air as big as beeches ever get and mantled in green
mosses; trees whose trunks have been pushed into a permanent north-easterly tilt by the prevailing
winds, and whose upper branches curve and strain back in the direction of this reconstructive force.
I know that the hurricane of 1987 did damage on these hills, and I do not have great expectations
that the trees survived in any number: they were beginning to fall even when I saw them. Aptly
enough, it is windy weather today, of the kind that 'may touch gale force on coasts and hills', and
rain is very close: the whole width of the westward sky is lined with ridges of advancing cloud.
Every few minutes the cloud thins a little to produce a lesser gloom, but swiftly darkens up again. It
is the most perfect set of conditions imaginable in which to touch the spirit of a hill like Lewesdon.

It is only when I draw near to the top of the hill that I become aware that the forecasters' pre-
diction has been realized: already the wind seems powerful enough here to push the trees over. For
some reason it is an exhilarating thought. The gale is roaring and wailing on all sides, its strongest
bursts making a sound like that of a storm at sea – complex, all-enveloping. The path I am using is
protected from the wind's main force by the shape of the hill; yet the wind is there, directly above
me, in the flailing and whipping top branches. It seems to be rearing up at a distance, threatening
death and destruction, yet here at least it passes harmlessly over my head. As the light dies behind
the thick cloud, the colour of the wood gradually becomes infused with amber, deepening to amber-
pink, and this colour itself seems to intensify as my eyes adapt to the gloom.

I find the beeches: the biggest of them stand at the head of a space on the shoulder of the hill.
Six or seven of them have gone over – though not today, as it happens – and lie now nourishing
fungi, although one at least continues living in its recumbent state, fed by a half root-plate that is
still linked productively to the ground. There are enough of the trees standing for me to know that

what I want from the hill, as a landscape, is still to be found here. Though it has many other beauties, these beeches are its defining element: one knows from them – by an all but unconscious comparison with memories of other beeches, similarly formed, cloaked and coloured – that one stands here on a West Country hill and nothing other.

I continue up the last steep slope to the top. This has one of those incongruous small populations of mature or dying Scots pines which have almost certainly been planted by an earlier landowner, wishing to improve his view in the fashion of the time. Remnants of such decorative pining are to be seen on quite a number of the central and western hills: on Martinsell, for example, above the Vale of Pewsey, on Pentridge Hill near Cranborne, and on the amusing little rounded knoll that stands not very far from here to the west of Bridport.

The summit of Lewesdon Hill is almost perfectly flat, much as if it had been sliced off like the top of an egg. It is possible to make out some faint traces of the Iron Age earthwork, whose construction did no doubt involve earth removal, not least because the area of the summit is very small. To the south-west, the wind is striking the steep ramp of the hill and planing upwards, blasting pieces of bracken and pale dead grasses towards my feet and face as I stand at the rim. I watch thousands of fallen leaves as they are snatched and gusted up this slope, taking flight and falling like flocks of sea birds. The upper parts of the great beeches – now below me – writhe and twist in the unrelenting current, while the smaller shrubs on the bank are driven into a sprung motion, jerking and dancing like souls in torment. This was not an afternoon to take pictures.

The next morning I return to the hill equipped for photography. Within minutes of my first attempt at a shot, low cloud has wrapped itself around the hill fort, shedding rain. The rain takes another three minutes to reach me on the lower slope, but when it does so it falls hard and steadily. I stand in the lee of one of the largest beeches and watch the first runnels of water beginning to deepen the colours in its bark. Tiny cataracts form at the lowest point of the curve of its branches and the water falls to the ground in needle-fine columns, now broken, now continuous. Gradually the bole of the tree is given a new, glossy and dimly shining surface resembling that of wet slate. It is hardly surprising that so many of the trees bear mosses. As I am cowering in this spot a jogger goes by, oblivious of my presence, in pursuit of her red setter. She is by no means slender, and the thought that she may have run up this hill for its full length from the north is impressive. Spurred into action, and using this dry spot as my base, I step out into the storm and continue working.

Like the woodlands of Ashcombe Bottom (see p. 45), Lewesdon Hill is one of those ancient woodland sites which – though its interest to naturalists and historians is not perhaps of the first order – has incalculable value both as a landscape in itself and as a feature of the wider landscape that it occupies. The wood can be seen from all sides, massed mysteriously on its summit like some dark, crouching beast in winter, or rimmed by the dazzling white blossom of the blackthorn in spring. It has inspired poems both by William Barnes (1801–86) and by the rather less well-known William Crowe (1745–1829), a resident of the village of Stoke Abbott, one mile distant.

It is reasonable to assume that since the hill housed a fort during the Iron Age, it was thoroughly cleared of trees then, which would give the woods that now cloak it to the north the status of ancient secondary woodland (see p. 85). The footpath to the south climbs the hill in parallel with another earthwork, a set of ditches and banks that run *down* the hill, across the contours. At some point in the past, these immense configurations were hedged in on both sides, and at the top of the earthwork, the hedge on its westward side turns at right-angles to continue on around the hill on a level. These hedges were planted with the same material that is to be found across central Devon and Exmoor, and on the top of the Quantocks, since it proved best adapted to their weather and soil conditions: beech. When management of the hedges ceased, individual plants grew out to their natural height on their earth bank foundations, which are still quite evident, and seeded themselves on both sides. Thus may a wood (or one part of a wood) spring into being out of nothing more than a set of former field boundaries.

Today, a path winds its way along the western bank, between the fanning growth of the smaller beeches. It is the kind of path such as is often made by that not uncommon woodland creature, the child: it leads nowhere in particular but does so in a delightful way, taking in every possible change in direction *en route*, and keeping down the growth of the moss in the process. As if in confirmation of my suspicions, at the northern end of this path I find a child's hide, or lair, cunningly constructed out of a lattice of broken branches, resting at one end on timber and at the other on the hedge bank.

The National Trust manages the woodland on Lewesdon Hill so as to encourage natural regeneration with a minimum of planting: the management plan includes the decision to remove all sycamores. In emulation of the wildwood, unless they block the footpath, the big beeches will be left to lie where they fall.

The rain moves in: a clearing on the southern slope of the hill.

A beech in the same location.

Detritus of the great wind of October 1987.

Outgrown remnants of a former beech hedge: the trees stand on the earth bank originally thrown up as part of the boundary.

A view westwards along the line of another former boundary.

April: looking south from the head of the chalk combe. The woods abut on former grazing land, and they are spreading outwards.

Ashcombe Bottom Woods

APRIL, JULY, OCTOBER

Privately owned mixed woodlands containing elm, oak, sycamore and ash, in which the last two are now dominant, with some hazel, thorn and dogwood, sited on both sides of a steep chalk combe in Cranborne Chase. The woods lie adjacent to a beauty spot, Win Green (National Trust), some five miles south-east of Shaftesbury, Dorset.

Apart from their established status as ancient woodlands that have not been replanted since at least the year 1600, the hanging woods of Ashcombe Bottom probably have no pressing interest for the landscape historian. They are typical of countless other similarly sited woods about the country, in that they occupy land steep enough not to have been recently required for agricultural use. In this stretch of chalk country, such a use would certainly have been sheep grazing, and though these woods are now being allowed to seed themselves outwards both above and below their earlier boundaries, it is not difficult to see where the former long-established clearance of the sloping upland and the narrow, curving combe bottom stopped. I chose to include these woods since they seem to me to illustrate well the value of many ancient woodlands as features of the *present* land-scape – as a cloaking of trees on hills that are most often seen not from within but from without. Like hundreds of other steep-slope woods, their presence transforms what would otherwise be merely beautiful (a long chalk combe, which if it were not wooded would most probably be under grazed downland turf, with some cultivation in the Bottom), into an Arcadian vista; one of those complex, mysterious views in which the elusive and often frustrating promise of the English landscape is realized.

The woods form a subsidiary part of a local beauty spot. People drive up to Win Green, the expanse of land on the hill above them with its solitary beech clump, to eat their Kentucky Frieds, or neck on the back seat, or indeed to stroll about on the top, admiring the 360-degree view from any point of the compass they choose. One of the best views southwards is to be had from a bend on the Ox Drove that runs along the chalk ridge: from here one can look out across the Bottom in the direction of Tollard Royal and the large remnant of the Cranborne Chase forest that lies east-wards of it. The combe drops away steeply at one's feet and the new spread of the woodland over what have been intervening pastures is all too evident: a battalion of young ash trees is densely massed here at the periphery of the mature woods, intent on marching up the slope. Like most chalk combes, this one is divided at its head into a series of subsidiary indentations. On the map it resembles an arm bearing a hand with open fingers, and each of the intervening spurs is capped with the soft, subtly rounded forms of the woodland, riding out on the broad and sweeping lines of the chalk.

I first discovered the Ashcombe Bottom woods while walking through the region one icy January day. It was a walk-for-walk's-sake – no cameras – which had earlier taken me through a long section of the Chase woods to the south-east. This was at a time when I had barely any understanding of the history of ancient woodland. Even so, there was one part of these combe woods, dominated by mature and over-mature trees, where the fallen timber had been left to lie and deer trails and droppings patterned the bare, muddy ground, and it impressed me with what I felt at the time to be an 'antediluvian' atmosphere.

During my most recent visit to the combe, on an afternoon in late April, there is a wind whose pummellings are strong enough to make my car shake on its suspension in the exposed hilltop car park. But the combe itself is all but windless, and as I descend the footpath from the Ox Drove (in 1992 still without either stile or sign), I move down in no more than a minute's walking into an almost trance-like realm of calm, in which not even the top branches of the trees are touched by wind. This path is little used, its course unclear at best at the start and wholly invisible lower down, where if you prefer you may follow any one of a dozen trails made by badgers or deer through the dog's mercury. At this time of year, the plant evenly and satisfyingly carpets the entire woodland floor, giving way only in a few places to wild garlic or stinging nettles. At the northern tip of the combe, the occasional Scots pine stands in isolation amongst the other trees. This apart, the only other suggestion of possible earlier management is an expanse of dead elm on an eastern slope. An entire block of trees has died here, leaving dense stands comprising trees other than elm on either side.

These woods are expanding downwards as well as upwards. The young ash has spread out along one bottom fringe, and where the trees are already well established, sycamore seedlings are also coming up beneath them. The 'tributary' combes to the west are floored with old pasture, long since abandoned – here it does take some effort to detect where the fences may have run – and the rank grassland is being colonized by thorn which will, no doubt, in time extend its invitation to the larger trees to join it. Along the south-facing side of one of these combes stands a strip of well-established young sycamores. The battle for supremacy in this part of the wood has long been lost to them by the oaks and elms, and whatever other trees may have grown with them: even so, and even here, the sycamores are not spreading at quite so alarming a rate as those who detest them would have us believe. There is something uniquely beautiful and poignant about a decayed farm landscape of this kind, in which – for whatever reasons, and for however long a period – all earlier human efforts at the control of nature have been abandoned. The one part of the combe now submitted to management is the central strip, running southwards along the Bottom, and even here, in the six months since I last visited it, there has been a change that many farmers would regard as a form of downgrading. In October, I was able to photograph the woods to the south over the stubble of a cornfield. Today, that cornfield is down to pasture, and freshly fenced; and the fence runs at quite a distance from the earlier boundary of the woods.

Not long after the sun has dropped down beyond the rim of the down, I re-enter the 'ante-diluvian' wood of memory and stand looking around at it, trying to work out what it was about it that so impressed me. Certainly today, under its even dog's mercury sward and with the tender new leaves just opening on low branches at eye level, it does not possess quite so powerful a mood as on the freezing winter day when I first saw it. But in essence it has not changed at all: the young trees stand rather higher, with the dead amongst them (the dead trees lie rotting just as they did), and after my many subsequent encounters with ancient woodlands I can see immediately why the image of the place should have stayed with me. It is precisely because this is a wood in which, as things stand, the only changes that occur are those that come about through natural causes. If anything can give us some sense or hint of the wildwood, it is surely this quality, which develops in any wood through nothing more dramatic than the cessation of human activity. Under the right circumstances it is no less present, I think, in a wood of over-mature sycamores on a chalk slope beneath a beauty spot than it is amongst a forest fragment of crinkle-crankle oaks on the Dartmoor clitter.

A section of elm wood killed off by Dutch Elm Disease during the early 1980s.

April: woodland edge. A deer trail through the dog's mercury, with self-seeded new growth of sycamore and ash beyond.

Detail of the root-plate of a fallen elm. This was once a fieldside tree in the combe bottom.

October: looking southwards along a cultivated section of the combe.

October, approaching sundown: the woods at the combe head with Win Green clump beyond.

Disease and malformation: one sure feature of the wildwood. The outgrowths here have probably been caused by insect infestation.

54

Pinnick Wood

EARLY NOVEMBER

One of the New Forest's 'Ancient and Ornamental Woods', an unenclosed oak- and holly-dominated woodland on wet land at the western end of the Forest near Ringwood, exhibiting all the classic characteristics of wood-pasture. A primary woodland, never cleared or replanted, and thus directly descended from the wildwood, though its structure has been modified by continuous grazing. It has a relatively rich ground flora which includes wild daffodil and some rarities, such as bastard balm, as well as an unusually rich display of bark-loving lichens.

To any streetwise but country-foolish fellow who harbours a dislike of wilderness, the approach to Pinnick Wood might prove more than a little disturbing, were he by some terrible error to find himself making the journey. Viewed from a car, the New Forest can seem a thoroughly urbanized landscape, little more than a wooded service area for the car-bound, peppered with picnic areas, camp sites and car parks prettily framed by trees, and traversed by single-track roads whose verges have been eroded away by the constant meeting and passing of traffic. Linford Bottom, below Pinnick Wood, is approached by just such a road, which metamorphoses at its northern end into a broad gravel track, leading to a yet broader car park. This is barred on all sides with substantial short wooden stakes designed to keep the four-wheelers and the caravanners off the Forest, and anyone who wishes to go on to Pinnick Wood must therefore get out and walk the remaining mile or so along the valley. Over this short distance, however, stage by stage one finds oneself abandoning civilization; first, by relinquishing the security of the car park; second, by walking alongside, and eventually parting ways with, the twentieth-century landscape of conifer-planted woods that lines the valley; and, third, by crossing the open heath until it begins to blur into woodland and the oaks of Pinnick Wood draw in around one. The final stage in one's progress into this isolated scion of the English wilderness will be taken only by the boldest and, joking aside, few enough seem to do it: this is to step off the paths altogether and traverse the woodland in any direction one chooses, like the ponies.

Yet for anyone who loves the wild, and is prepared to stand a while and contemplate the mood of a place, Pinnick Wood has about it a palpable benignity. There is something profoundly reassuring in the silent presence of all these big old trees, and in late autumn the wood has a scent that is as sweet and clinging in its way as any fine Tibetan incense. Only a few of the trees here are very old: there are fewer than 25 oaks of more than 350 years of age in the whole of the New Forest. But the wood itself is ancient, and has probably stood this ground, changing in composition and later in size and shape as it did so, since the first post-glacial warm period some 10,000 years ago, when trees first began to colonize Britain from the Continent.

What one sees here today is an oak wood possessed of a substantial holly understorey. Many of the hollies are massive; and some of them grow in random pairings with individual oaks, standing with their branches upthrust or nearly intertwined like dance partners frozen in mid-motion. In common with all the other intact 'Ancient and Ornamental Woods', Pinnick is free from the decayed trappings of medieval or later wood-management. It is unenclosed and has no boundary banks: hence its indeterminate margin. Also there is no sign of coppicing here. The animals — deer, cattle and pigs, as well as horses — are able to range freely into and through it, and it seems that this has always been the case. The wreckage of fallen trees is left to rot and there are none of those other signs of tidying up, such as sawn-off branches (let alone wood piles or hard pans for lorries), which one finds in modern managed woods. The oldest trees, too, are left to die where they stand: in Shakespeare's day these went by the appealing name of 'doddards', which, had there been pollarding here, it would be tempting to regard as a portmanteau word. It is no coincidence, then, that the 'A & O' woods possess an abundance of species that are dependent upon dead or dying wood, or that take up residence in holes in trees. For example, 13 out of Britain's 15 species of bat are to be seen in such woods. Just as in the pine forest of Eilean Suibhainn (see p. 155) on Loch Maree, young trees stand here next to the old, dying or dead, whose claw-like forms could not contrast more strongly with their own straight boles.

Yet for all this, can Pinnick Wood truly be said to be free from the influences of the twentieth century? As I traverse its branch-littered and bracken-patched floor in search of shots — this is not difficult, since there is plenty of space around the obstacles — I keep returning to the broad paths, where any hoof and trotter prints have been entirely overlaid by continuous patterns threaded through the mud by trainers and boots. On one of these paths I encounter a pair of elderly ladies, walking Labradors. The monstrous A31, which slices the New Forest into two disconnected halves, a northern and southern, runs less than a mile to the south of this 'primeval' place: its roar is constant. Distantly and repeatedly, too, from the army playgrounds to the north, comes a sound like a slamming of doors, which somehow finds its echo in the humus beneath one's feet, in the branches above one's head, in one's own heart, and seems to send the acorns tumbling and scattering a little earlier than they might do otherwise.

One of these hits my camera on its tripod with a tiny impact. I hear a disturbance behind me; but when I turn there is nothing there, not even a leaf-shuffling blackbird. On my first visit, my onward progress through the wood is no more systematic than that of any animal: I graze it for shots in easy pursuit of its countless sculptural forms, in the process moving from one very temporary encampment to the next. At any distance from the paths I find that tripod, camera-bag and rucksack gradually become separated from one another across the woodland floor. These objects are markers of my 'possession' of the place, for the duration of their stay, although, of course, true possession in such places comes only when one is least expecting it, in the shape of a heightened perception. And it is then too — when one perceives the place with or without the help of lenses — that it becomes wholly peaceful. At that point, for as long as the feeling lasts, the manic road-roarings and the thumpings of the artillery cease to have any currency or influence here.

There is another curious noise. I turn to see a small congregation of half-grown pigs nuzzling their way through the fallen oak leaves, pinker than rock-candy among this landscape of muted russets and deeply shaded greens. I begin to trail them — this is a shot that I *must have*; is this not the living embodiment of *pannage*? — but they know perfectly well that I am after them and refuse to stop and graze at any useful distance, or in any useful spot. As I walk on behind them, increasingly frustrated, the wood-floor seems suddenly full of dry and loudly cracking twigs, and I lose sight of the first of my encampments, where most of my lenses are stored. So I turn and go back. Knowing that I have lost interest, the pigs gradually re-form in the area where I first saw them.

It is possible to understand something of the meaning of the phrase 'primary woodland' in a wood such as this. As always, it is the site that is of the first importance, rather than the individual trees which happen to be growing on it now. A wood such as Pinnick has not only never been cleared for farming; it has never been submitted to coppice management, drained or replanted. However, it has been continually *grazed*: hence its status as wood-pasture, a term that is applied to all the unenclosed parts of the New Forest as a whole. The predilections of grazing beasts have very slowly, but also very distinctly, altered Pinnick's species composition. Pollen cores taken from valley bogs have shown that the primeval forest in this region was dominated by small-leaved lime, accompanied by oak, elm and hazel, and by alder in the wettest parts. Today, there is no trace of lime, elm or hazel in Pinnick Wood: they have been browsed out of a lineage here. The grazing was, in a sense, selectively damaging to the wood. It did not kill it; instead, over time, it readjusted its structure to the long-standing advantage of the oaks. This is not to say that grazing pressures have never been excessive here. In 1850 – a period of prehistory, of course, when measured against the time-scale of modern environmental management – the New Forest Commissioners came to the conclusion that the pressure of grazing from the deer was great enough to prevent regeneration across the Forest as a whole. Legislation was swiftly obtained – it took only a year – and the deer were excluded over an area of 4,000 hectares by the use of temporary fences. This resulted in a wave of new growth in the enclosures, and many of the 'A & O' woods also increased in area.

It may not be obvious to the casual visitor, but Pinnick Wood contains many other lineaments of the wildwood, which combine into what naturalists call a 'complex ecology'. Quite apart from the bats, the wood can boast an outstanding population of insects and other invertebrates, and is rich in bark-living lichens, including some that can only survive in woods that have never been coppiced or clear-felled. To trace and isolate a few such species is certainly to come a little closer to grasping the antiquity of such a woodland.

At the end of my last visit to Pinnick Wood, I found that I simply did not want to leave it. Nor did I want to photograph it either. Such was my timetable that I had little choice in the matter, although I did allow myself to sit for ten or fifteen minutes, no more, beneath a particularly striking tree. A few months later on my travels, I happened to meet a man who had himself recently spent some time in the New Forest. He walked with purpose, and in general had the look of one who knew what he was about. He told me that he had just spent six weeks camping in the Forest, watching the phases of the spring as it took hold there. During this time, except to visit the occasional shop, he had never left the woods – another way of relating to nature besides that of photographer, huntsman, and so forth. I am afraid that it did not occur to me to ask him whether he had identified any specially interesting invertebrates.

There is no man-made boundary to Pinnick Wood. New Forest ponies and other animals graze through it freely.

A group of oaks near the wood's north-western margin.

Off the paths: dead and living trees at the centre of the wood. Most have been prostrate for many years.

Skeletal remnant of a long-fallen oak.

The dance of the trees: one of several bizarre pairings of oaks with massive hollies.

Regeneration of the ancient: where the wind tears away a branch, the yew will sometimes send out shoots direct from its trunk.

Kingley Vale

NOVEMBER AND LATE APRIL

A National Nature Reserve, managed and partly owned by English Nature. Regarded as the
finest yew wood in Europe, the reserve is sited over 150 hectares in a combe and on the slopes
of an outlier of the South Downs, four miles to the north-west of Chichester. Most of the
reserve is occupied by a recent secondary woodland in which the trees are between 70 and 100
years old; but there is a central stand of about 20 ancient trees growing on a deep bed of combe
rock, and these are thought to be more than 500 years old.

In the nineteenth century, the old yews at Kingley Vale were given the nickname 'the Druids' trees'.
Like another legend attached to them – that they were planted here to commemorate a battle fought
and won against the Vikings in AD 859 – the association is as spurious as it is attractive. Yet this is
surely one of the most fantastic stands of old trees in the country, and if ever a theatre designer
needed material for a backdrop to some hypothesized fantasy of Druidic ritual, he would need to
look no further than Kingley Vale for inspiration. To walk under these trees is rather like penetrating a series of interlinked caves, each of whose roofs is supported on a massive, noduled and mis-
shapen pillar. Though all around them lies an impenetrable weave of bramble, elder, dogwood and
holly, overlaid in places by the white drapery of old man's beard, in the shade of the trees nothing
grows: the cave floor is bare, save for needles and bark discarded by the trees themselves.

On my first visit here I follow the nature trail, as do most other visitors, from numbered post to
numbered post, and it leads me on through this string of darkened spaces in which there hangs a
sweet, fusty perfume from the fallen and rotting berries. It is impossible, surely, not to read into the
liquid and rippling folds of these ancient trunks some parallel with animal musculature; equally
impossible not to see their clefts and cavities as distended, exclaiming mouths. There is in each the
look of some bizarre and hairless beast whose deformed body arches and strains upwards towards
the sky, this thrust continuing on into the branches before they themselves fold back down again
landward, as if dragged there by the weight of their massed dark foliage.

The '87 wind did some damage in Kingley Vale, but it seems that only one of the ancient yews
went over. Instead, the wind tore down the heaviest branches from some of the trees, in the
process splitting them open and letting in light for what may have been the first time in centuries.

One great tree at the periphery of the dark central 'passage' stands now with its branches wrenched outwards, fallen away from one another. The red heartwood projects raggedly from the centre of this massive bouquet, while a lesser tree lies cradled across it. Another yew has split down the centre of its trunk and survives now as, in effect, two separate trees: this is not the first time that such an event has taken place here. On some trees, where a branch has fallen, fresh branchlets now appear in clusters, growing directly out of the surface of the trunk: they resemble tiny bushes stranded on an inhospitable cliff-face.

Here once again is a collection of ancient trees that may perhaps allow us to form some mental picture of the appearance of primeval forest. However, it is well worth bearing in mind that although the trees are indeed very old, they do not in themselves get us much further back in time than the late Middle Ages, and no pollen analysis has yet been carried out here to establish if there is a clear connection between the site and earlier times. Remnants of field systems have been found, showing that the combe was farmed by the Saxons, and it is perfectly possible that the 'Druids' trees' sprang up from seed transported by birds in just the same way that the more recent forest has done, on land that had earlier seen many centuries of service in human hands. As is often the case with ancient woodlands, one may go to a place such as Kingley Vale to fire the imagination; but there are practical limits to the heat that is to be had from the blaze, at least if one is to approach the subject with any degree of realism.

This Nature Reserve is a popular place, well known to the Sussex people, and its location three-quarters of a mile up a track padlocked against vehicles does not seem to keep visitors at bay. It is equipped with a little gazebo-like museum, stinking sweetly of formalin, which acts as a mausoleum for the preserved corpses of birds and animals, including a sparrowhawk, a baby deer and the mask of a badger, each of which has been found killed on local roads. There is a nature trail leaflet, which visitors can use to guide themselves around, following the posts leading to laminated interpretation boards. Near the first of the big yews, a garden chair of antique design has been attached to a low branch with a cycle padlock. It is pleasant to imagine that it might have been left here by some regular elderly visitor, and perhaps used while he or she sat dabbing paint on canvas. But most of the wooden slats in its seat have long since rotted through.

Beneath the old yews themselves the trail is worn in: the earth shows up dark where needles and russet fallen bark have been scuffed aside, and the trees' roots are glossy where countless feet have polished them; so too the trunks, anywhere near the central trail, which have gradually lost their bark under the touch of inquisitive or grasping fingers. In the hollow of one great stump I find a few crumpled sheets of pink toilet paper. Side-paths have been driven into the scrub between the trees where the nature trail leaflet does not encourage them. To step into any one of these culs-de sac in the steps of the pioneers who made them is instantly to confront the danger of true forest: it is almost impossible to keep any orientation, save that of remembering how to retrace one's steps. The great beasts – the yews – loom up here out of the dense web of bramble, quite unexpected, barely seen until one is upon them, waiting presences if ever any were. The place may be popular, but nothing short of a coachload of football fans armed with ghetto-blasters could completely dispel its mood.

There is very little light under the yews. Even at two-thirty on an afternoon in November, I am working with one- and two-second exposures. Later, the shade is so deep that I know I shall see the full detail of what I am recording only when I view the transparencies. My first afternoon's work here is conducted in an all-too-familiar state of suppressed and needless panic, although the fear is even less rational than usual: not that the light will be gone before I have achieved anything, but that the subjects themselves will in some way lose the very qualities that make them interesting, never to regain them.

Today I have been working without a tape measure, collecting details of the trees' tormented 'musculature' and 'mouths'. When I have the films back, I find that most of these shots are quite useless, since only a part of each image is sharply focused. It is a pity: they are, or would have been, the best of the lot. I take the dud transparencies with me when I return the following spring, since I know all too well that it is difficult to repeat a shot in woods unless one has the clearest point of reference from which to work. But I am not able to match a single one of them. No matter how hard I look, the conformations I have on film will not correspond with the actuality, except in two cases, and here the light of a different season makes the subjects uninteresting. Of course, it is quite clear what has happened. In the five months between November and April the trees have moved their arms.

Details of the boles of two trees. Both are still living.

The 'Druids' trees': near the centre of the most extensive stand of ancient trees.

Nothing can live in the shade beneath the biggest trees . . .

. . . Bramble encroaches on the bare ground only where the wind damage lets in the light.

Where the grass grows greenest: the beginnings of a watercourse.

Binswood

Early November

Sixty-two hectares of wood-pasture common on heavy acid soils, owned by the Woodland Trust since 1985, and situated four miles south-east of Alton, Hampshire. Once part of Woolmer Forest, it was established as 'folkland' by Henry II, but was subsequently used as a royal deer park by King John before being re-established as common by Henry IV; it may still be grazed today by those with rights to common, and is equipped with a 'common warden', who now reports to the Woodland Trust. It consists of stands of parkland trees, mainly oaks, and areas of denser woodland in which oaks, beeches and hazel coppice predominate; a very few old oak pollards also survive. Notable for acid grassland plants such as bitter vetch, tormentil and lousewort, as well as for its wide variety of butterflies.

The first and most enduring impression of Binswood is that it is a secret place, untouched by any of the roads around it, invisible from them, and accessible only by way of bridleways or paths crossing the buffer zone of fields that frame it. It is partly this separateness from the modern road-world which has enabled it to keep its atmosphere. Certainly, as one climbs over a stile and passes the Woodland Trust sign for the first time, one has the most vivid sense of coming into a new kind of landscape, in other words (and the two are inseparable) of confronting another concept of land use than those that are generally visible in contemporary England. For this is, or ought to be, managed woodland that is also grazed, grazing land that is also productive woodland.

There is some danger of making too much of this, as the Woodland Trust itself has done when it claims in its leaflet on Binswood that it is 'the last working wood-pasture common in Britain'. This is rather to overlook the 3,671 hectares of unenclosed ancient woodland in the New Forest, which also conform to Dr Rackham's definition of wood-pasture as 'tree-land on which farm animals or deer are systematically grazed' – at least, if one interprets the word 'systematically' with a certain freedom. This said, however, Binswood remains very much its own place. Seen from the north it impresses first as a landscape in which all the distant groups of big oaks might have been deliberately planted (some, but by no means all, were). These trees stand together in threes and fours, catching the sunlight of a clear November morning in their leaves; lost amongst the bleached-out grasses of an overgrown pasture, which are collapsing in on themselves as the autumn ends. On a day like this, these groupings strongly resemble the groves of some Claudian idyll, and as I follow their magical, backlit 'autumnall' southwards, I find myself moving through a kind of oaken savannah, dotted with a mix of individual trees and larger clumps, yet still very much open to the sky. Eventually this treescape thickens into the large area of woodland that covers most of the southwestern part of the common: this is pleasing enough in itself, but by no means as beautiful as what stands around it. Other approaches to Binswood allow variations on the same experience: from the west, for example, the walker comes into a finger of open land framed on both sides by the denser woodland; approaching from the east, he enters a broad circle of pasture and groves, beyond which the woodland lies. It is in this second area that you may see some of the remaining oak pollards, although most of them are now dead.

It is always difficult to photograph woodland on a sunny day. The stronger the sun, the blacker the shadows (at least, when one is shooting on colour film), and the ideal moment usually comes just as a cloud is about to move on past the sun. While the cloud is still there, the world looks dull and lacks contrast; and when the sunlight is fully released, back spring the impossible blacks. For a very few seconds between these two, when the sun is shining through the rapidly thinning cloud edge, there is a moment of perfection, but to catch it one must wait, and wait, stalking a few seconds of time with one's camera – again, the parallel is with hunting. I do so today, making half a dozen exposures across each transition and, as often as not, still somehow managing to miss the essence of the thing. Clearly I should be shooting the oaks with a motor drive, treating them like VIPs running the gauntlet of journalists from their Mercedes, or athletes clearing hurdles. Even so, my periods of waiting are not without reward. The smell of the late autumn is everywhere in this unusual warmth: always it seems to be hinting at some entirely non-enforceable promise of fertility and the richness of the land in the year to follow. It is at once sweet and acrid, this smell, rising from the ground where the oak leaves, the fungi, the grasses, the fallen nuts and fruit are all beginning their decay down to humus, as vigorous to the nose as the smell of fermenting wine.

Binswood is an ancient site, but in contrast with the best of the New Forest woods, it is difficult here to see any direct connection with the wildwood, unless it be by way of its collection of lichens. What one does see is a fragment of land on which a medieval system of wood management was applied without interruption, or apparently a great deal of modification, until well into the present century. From the time of Henry IV, the arrangement was that commoners had the right to graze their cattle or other livestock in Binswood during the summer months, thus freeing their own small fields for the production of winter feed.

There are still twelve commoners today, at least in name: the rights of grazing are passed on with the freehold to their properties. However, the late twentieth-century commoner tends to have rather wider-ranging interests than his predecessors, and other sources of income than a couple of two-acre fields. It is hardly surprising that the Woodland Trust should be finding it difficult to persuade many of the new incumbents to take up the opportunities that are open to them. One of the commoners is a full-time farmer, who, as a result of a reorganization of rights some 30 years ago, may if he wishes graze a herd of 50 cattle on Binswood. In earlier years he has done so, but in 1991 and 1992, in a reaction to the BSE scare and the depressed price of beef, he made no use of the common. The spread of bracken – which carries a cattle disease called red-water fever – is another disincentive, but the Trust is now beginning to gain control of the plant. Ironically, although it is the owner of the site, the Woodland Trust itself has no grazing rights. The restoration of the pastures (and, indeed, the understorey of the denser woodland) at Binswood thus remains a problem, though in the long term it is not an insoluble one; in the short term, it may well be necessary to bring in a tractor and do a little mowing.

In direct contrast with the New Forest woods, Binswood contains large areas of hazel coppice, which could have survived the pressure from cattle only by being fenced off during each period of regrowth. In 1991 there was one well-constructed deadwood fence at the centre of the common, serving exactly this purpose around a triangle of recent coppicing: the threat at present comes from the deer. The thinning of the coppice will continue – this is one aspect of management that can be carried out irrespective of the involvement of the commoners.

The currently overgrown state of the Binswood pastures is all the more ironic since, until the Trust's purchase of the land, it was suffering as a result of being heavily overgrazed by its owner.

The one animal to be seen grazing in Binswood. But note the electric fence.

Oaks and fallen branches in the northern reach of the pasture.

In the densely wooded central area of the common: just before dark.

A view out towards the eastern grazing land.

Near the eastern margin: oaks and undergrazed pasture land.

October: plantation trees, typically without lower branches, near the centre of the wood.

Gutteridge's Wood

OCTOBER AND EARLY NOVEMBER

Beech plantation on an ancient secondary woodland site, with some oak (planted) and yew
(probably self-sown), matured into high forest on chalk, some five miles north-west of Reading,
Berkshire. It is typical of many Chilterns beech woods managed for their timber. (Gutteridge's
Wood is one part of an interlinked series of woods including Hawhill Wood and Holme Copse,
where photographs were also taken.)

In the introduction to its inventories of the country's ancient woodlands, English Nature (or the
Nature Conservancy Council as it was known at the time) made a fundamental distinction between
what are termed 'ancient primary' and 'ancient secondary' woodlands. For a wood to be classed as
'ancient' at all in the survey, it was necessary for it to be standing on a site that had had continuous
forest cover since at least the year 1600, which is the time when plantation forestry first began to
be taken up in England on a large scale. As in the case of Pinnick Wood (see p. 55), primary ancient
woods are those that allow us the most direct connection through time with the wildwood, since (it
is known or believed) they have never been cleared for farming. Secondary sites are those that
were thus cleared at one time or another in the distant past, but have been subsequently re-wooded,
either by deliberate planting or through natural regeneration, before the year 1600. Many of the
Chilterns beech woods would appear to come into this second group, and even if they are in fact
much more ancient, their present composition will usually frustrate any attempt to establish an earlier
history for them.

The difficulty with dating the Chilterns woods lies in the fact that the beech cover has stood here as
high forest over a period of at least two centuries. Trees were taken out using the Selection System – a
term that is self-explanatory – and woods were therefore rarely clear-felled. This continuity of beech
cover has had the effect of suppressing other vegetation, since the beech's surface roots remove much
of the moisture from the soil, causing other species to die of drought. The tree's leaf mould is also inhospitable to many species. Generally, then, it is difficult to judge the true age of a beech site simply by
examining it for ancient woodland 'marker' plants. We know that in the seventeenth century almost all
today's beech woods occupied the same sites, or near approximations of them. We also know that
though some beech woods were planted for coppicing during the same century – mainly as a source of
fuel for London and the towns of the Vale of Aylesbury, by then largely deforested itself – there were
others that *already* stood here in the form of coppice when the planting began. It does not take much

imagination therefore to begin to speculate on the possibility that for all their present ecological simplicity, some of today's plantation woods may none the less be standing on primary sites.

The beech woods are frequently dismissed by naturalists as being of little ecological interest. I am in no position to argue the case in such terms – although I would say in passing that I have seen more than one expanse of wood anemone, wood melick and spurge living with apparent ease under the beech, and I know a number of plantation woods, so-called, that possess mysterious understoreys of holly, or stands of yew. However, from the point of view of the photographer with an interest in abstract forms, the Chilterns woods have few rivals. Those concerned solely with a pursuit of the wildwood may well go first to other woods, but if they have any feeling for the beauty of woodlands as a thing in itself, then they cannot ignore the beech woods.

Gutteridge's Wood runs up to the north of a busy back lane, a rat-run between the fair town of Reading and outlying suburban blocks and villages. The wood's margin is unfenced where it touches this road; a pub backs on to it, and people like to park here and wander into it from any angle. The standard markers of urban proximity – old crisp packets, beer cans, rank piles of dumped bedding, black bags – bespatter this boundary, while a little deeper in at the western edge there lies a burned-out car, with leaves floating on a pool in its dented roof. Nearby on the ground is an electrical coil: it looks usable. Yet walk another 50 yards further and the quality of the place begins quietly to weight itself against such bagatelles; and as it does so, the world turns about. Glimpsed from the road, from a car window, the wood is just another anonymous tree-fragment decorating the urgent abstract of bends and passing-places on the route, flashing by in a matter of seconds. But from within, as one looks back through the trees, it is the road with its busy, gliding cars that seems limited, sealed into itself. Now it is the *wood*, 'plantation' wood that it is or may be, which looks as if it might go on in every direction, rolling away over the Chilterns as forest without end.

What makes the beech so beautiful when seen in numbers is precisely the fact that – as a friend who was completely new to the woods once said to me – at first the trees 'all look the same'. Yet within this apparent homogeneity there lie as many formal variations as the mind cares to uncover. In reality, of course, no two beeches are the same; no more than any two of the flints lying between their roots are the same. The variations here are of an infinite complexity, yet they exist within a narrow band of reference. They may best be perceived through photography, at those times when it is used as a method of hard leverage: the imposition of the rectangular or the square frame can give sudden definition to patterns that are grasped only superficially by the wandering eye. For observation alone to be effective in such places, one's powers of observation – contemplation – would need to be strong indeed. Granted, the raw materials of these variations are inherently beautiful. In overcast light, the liquid sinuosities of the darkened trunks make them look almost as if they had been poured from above, or had run down like spilt paint on the backdrop of light-retaining leaves. In the first and last light, when it hits them directly, the trunks become incandescent: there is a period of 10 to 15 minutes in which they will glow as if illuminated from within, and this illusion grows stronger the further back the trees stand in what is otherwise deep shade.

At the end of the day, when I am shooting at the edge of the wood, I find myself penetrating the spaces in front of me, probing them almost as if I am staring into a partly darkened room, and using a long lens to isolate details that are remote to the naked eye. Repeatedly at such times I find myself drawn in deeper amongst the trees, and then again deeper. Always there are trees further on, where the shadow-fall and the juxtaposition of glowing shapes seems 'more perfect'. So I walk forward, and unless I am very careful I may lose the angle or, for that matter, lose my intended subject altogether. In these last few minutes of direct light, the shadow of one tree will move across another so quickly that a composition, or what makes a composition interesting, can be lost in the time it takes to change a lens. So, too, a glinting scatter of autumn leaves may be thrown down by the wind before, during, or after the moment when the trigger is released – it is just a matter of timing.

All too often, and especially with landscapes, a photograph will create the illusion of timelessness in its presentation of its subject. Yet to the individual who has been involved in producing it, the work in itself can reveal precisely the opposite: the mutability of all that we find around us, and the process of mutation identified, exposure by exposure, *in medias res*. More than once in Gutteridge's Wood I have forgotten my shot altogether, distracted into 'watching' the shadows move – one does not see the phenomenon, not quite – and feeling in their silent progress a sudden, dizzying connection with the larger progress of the earth around the sun.

For all this, and even before they have been immobilized by photography, the beech woods do have about them a quality of timelessness that can be very difficult to see through. When one enters a wood for the first time, it can take all one's mental effort not to imagine that it has always been like this – *exactly* like this – and will continue to be so, at least (one has to say) for as long as there is no further dramatic deterioration in the outside influences on its health. The effects of great winds aside, it takes years of looking at the woods to recognize concretely for oneself that they exist in a constant state of flux, and would do so even if no felling ever occurred. Even if that were to be the case, trees would still fall of their own accord, gaps would appear, and the new growth that filled them here would as likely be birch, or ash, as beech. A beech wood left alone would not be a beech wood pure and simple for very long. It is our separateness from nature that engenders this dangerous illusion of changelessness, and I think the photographic image plays more than a peripheral role in feeding and endorsing it. The grief which resulted from the hurricane damage of the late 1980s is evidence of the commonness of this state of mind. It seemed that what the bereaved had most wanted was for their wooded landscapes to remain fixed precisely as they were, as if in paintings, or photographs.

Gutteridge's Wood is in my own stretch of country: on and off over several years I have spent time there and in the neighbouring beech woods, photographing the trees, exclusively in black and white. There was a time when I might have been tempted to claim this wood in particular as 'my own': had I not demonstrably come to know it, through photography, as did no other? But on the very last of this November's forays I find myself leaving the wood at ten o'clock and returning to my car on the roadside. While I am fumbling for the keys, another car pulls up twenty yards in front of me. While I am packing away my equipment, this car's driver is getting out and intently removing his. He seems wholly oblivious of my presence; or perhaps it is just some unspoken decorum that keeps him from acknowledging the proximity of another camera-user. As I am driving away, this rival – this man who seeks to challenge *my* relationship with the place – is striding off into the depths of the wood. It embraces him among its golden leaves just as two hours earlier it embraced me, and he goes on until I can no longer see him, scouring it for shots – new shots – in just the way that I did. I content myself with the thought that at least we did not come to blows – as we might have done, perhaps, had he arrived half an hour earlier – over where precisely we should set up our respective tripods.

Even in plantation woods some young trees will seed themselves.

The very last minutes of direct sunlight, striking beech trunks at a west-facing boundary.

In October, direct light from the low sun enters the wood selectively, as if from a series of moving spotlights.

November: fog.

October: pink dusk light and a path at the wood's western end.

Dead and living oak pollards.

Staverton Park

LATE MARCH

Fifty-five hectares of privately owned woodland harbouring some 4,000 ancient pollard oaks, with hollies and birch, established on very acid blown sand seven miles to the east of Woodbridge, Suffolk. One of the best preserved medieval parks, managed as wood-pasture in the Middle Ages for both timber and grazing. Poor in flowering plants, but both and oaks and hollies support lichens whose presence suggests that the park may have been created out of primary woodland.

Here, if anywhere, in our pursuit of the wildwood we touch upon an enchanted land. Yet the contrast could not be greater between this place, with its unimaginable display of weirdly shaped oaks, massive hollies and ribbed and green-stained birches, and the scraped-naked agribusiness landscapes and desolate forestry plantations that butt up next to it. To the east lies a good working example of Suffolk's sandy carrot and wheat-surplus country, where straight blocks of conifers patch long views of nothing in particular, and polythene sheeting lies white across the fields like dead water. To the north and west is the orderly spread of Rendlesham Forest, its tens of thousands of hurricane-blown pines now bulldozed into half-mile-long piles of waste, with new plantings ruled on the land in between. Yet come up to Staverton Park on the B-road from Butley and suddenly along the hedge-starved verges you see the first old oaks, isolated against the void like harbingers of some entirely different kind of landscape and, indeed, of some other and incomparably richer concept of land use.

Enter the park as I did, along its one public path, and within half a minute's walking you are lost amongst the gigantic hollies of its southern part, The Thicks. The largest holly in Britain is reputed to stand here, somewhere – I did not find it – and there are many others of more than 150 years of age. Dare to wander from the path in this wood and – unless you have a compass with you, or can take a bearing on the sun – you will be lost in minutes. There is no view through the trees here: all views are completely blocked by the shining low-branched holly, and where trees have fallen, the bracken climbs up and clings about their decaying branches as if to double the difficulty of the obstacle. Deer scatter in the deep shadows, though not at any great speed: they know that they are difficult to see.

Continue along this path for another five minutes, however, and the wooded landscape will open up before you. Here, to the west of the path, lies half a square mile or so of land on which are spaced the thousands of ancient pollard oaks for which Staverton is famous. Most of these trees stand wide apart on a floor of rank grass and bracken, so that from every angle it is possible to enjoy long vistas between them. And what vistas these are! Each tree is its own creature, bulbous-trunked, alligator-skinned, with its branches thrust high from the top of its trunk like rigidified tentacles. Any one of them might have made a fitting subject for a landscapist such as John Crome of Norwich. Some stand fixed in a lean, or curve out drunkenly, overhanging space; some are conical, some cylindrical; some resemble the stumps or stubs of much taller trees, now unnaturally fattened. Some have cloven trunks, while others are much further down the path to decay, their opening wounds spewing a brown-black substance that is already more earth than wood. In the last stages, when the tree has already split open and fallen apart, its hollowed remnants form curving window-boxes of humus in which ferns and even birches take root. The trees' history as pollards is written all too clearly in their forms, in other words. Almost all their branches grow from points 15 to 20 feet above the ground, well above browsing height, though cattle that might have threatened the new shoots have not passed between these trees for centuries.

Here again is some hypothetical correlative of the kind of harmonious disarray to be found in the wildwood, standing its ground in present times. Yet it is to be found in a woodland that has only the most tenuous of connections with the thing itself. Staverton Park was a medieval deer park, founded some time between the arrival of the Normans and the middle of the thirteenth century and owned by the Bigods, Earls of Norfolk, although it is possible that even by the late years of the thirteenth century it was no longer used for deer. Surviving accounts from this period, discovered by Dr Oliver Rackham, show that the park was managed at that time for its timber (including sales of bark from felled oaks, used for tanning), for grazing and *pannage*, and for the sale of the cut bracken, which was used in those times as a litter for stables as well as being burned to produce ash for glass-making. The oaks were probably last pollarded towards the end of the eighteenth century, and since they vary in age from 200 to well over 400 years, we can be fairly sure that what we are looking at here is a gone-to-seed form of wood-pasture, some of whose trees date from the sixteenth or possibly the late fifteenth century. The natural continuity with the wildwood exists here only in the presence of the lichens, which suggests that the original park was adapted from an existing stand of primary woodland.

On my first afternoon in Staverton Park, for more than an hour I find myself quite at a loss. Where to begin, in the abundance of strange 'wild' beauty that confronts me? As usual I make a series of nomadic encampments, nibbling timidly at the display like some kind of visual herbivore, a little here, a little there. It is more than an hour before I get into my stride, and begin to *see* the place.

A gale is blowing, and all around me there is a roaring and squealing of rubbed branches. Then, quite unexpectedly, another noise penetrates the gale: it is a whining scream, machine-made but not instantly identifiable, and it is followed by a shuddering that seems to take hold of earth and sky together, and shake them. Three minutes later, one after another, a series of black flashing points of energy skim the sky above the thin tree belts to the east. They are fighter jets, and – of course, I had completely forgotten about it – they have just taken off from one of the runways at Bentwaters Aerodrome, one mile to the north of Staverton Park. The choking, sickly scent of aircraft fuel drifts towards me through the trees, and a disembodied tannoy-voice emerges faintly out of the wind, almost as if its source were the trees themselves. Now the park is quite another place than it was three minutes earlier. With all their rootedness equally in space and time, the great oaks remain at centre stage; but just in the wings of this stage lies another and powerfully qualifying presence. There is a threat in the air: it is unspecific, indecipherable, but quite impossible to ignore.

At four o'clock a tinny pre-recorded blast of the national anthems of, respectively, the United Kingdom and the United States reaches me through the oaks' mad branches, while the trees stand, grey-brown, fissured, ancient, motionless only at their uppermost tips, each possessed of its own centre of quiet. They seem to me (but I am partisan) like some symbol of the much shrunken realm of the imagination, upthrust in protest against that other and far larger realm that lies on all sides here, in which imagination has been declared entirely surplus to requirements.

A massive pollard. Its profuse branch formation is unusual even in Staverton Park.

An oak with a very large birch, typical of the site, beyond it.

Looking north, towards Bentwaters Aerodrome.

Living trees with hollow trunks and crazy, tangling upper branches:
the norm rather than the exception here.

Beeches above the cliff edge, bearing the first tender open leaves of Spring.

Lady Park Wood

LATE APRIL

A National Nature Reserve of 44 hectares, owned by the Forestry Commission and managed by English Nature, situated on high land to the south of the River Wye, three miles upstream of Monmouth, Gwent. A mixed high forest of ash, beech, hornbeam, lime, oak, maple and yew, among other species, with old hazel coppice, it is one of the most important sites in the Wye Valley system of native woodlands, which collectively contain a greater variety of trees and shrubs than any other comparable area in Britain, and are regarded as of international significance.

Certainly the seeker after ancient forests may bear Lady Park Wood in mind, or conjure it in his dreams, or for that matter go to the length of walking the rounds of its fenced perimeter on public footpaths and peering into it from different angles. But unless he has valid research to conduct, he will not be allowed inside it. This is one of those woods in which human interference of any kind is being eliminated as the basis of a long-term study in the behaviour of their plant life once management has ceased in them. The cover here has developed into high forest over the past century, which means that the naturalists already have a clear running start for their study: there has been very little human activity in the wood during this time. It is considered by some that the broad mix of species to be found here may already present us with a convincing picture of the ancient indigenous forest in this part of the island.

If Lady Park Wood does hold a mirror to the wildwood, then in its present state it seems a place of uncommon spaciousness, even having about it a certain airy elegance, at least in late spring. There are areas of old coppiced hazel here and some gaps have been made in the canopy, by wind-throw, in which young trees are now regenerating in dense stands. But in its central area directly above the Wye cliff this is a woodland of large trees, some of them magnificent specimens whose lowest branches project 50 feet or more above one's head. There is little need here to force one's way through an impenetrable mesh of low branches, and so orientation is easy. There is some fallen timber, but it does not yet litter the woodland floor in the kind of grotesque chaos to be found, for example, in Pinnick Wood (see p. 55). Occasionally a group of oaks with lion-limb roots will rear up cleanly, as if they have had few obstacles to a healthy and swift growth: of course, this makes them all but impossible to photograph. In the hazel coppice in the higher part of the wood, fallen birches are gradually sinking into insubstantial, rotted-through images of their former selves amongst the leaf mould, and there is much ivy here, of the most vigorous and potentially destructive kind. In more than one place it has overwhelmed a stem of hazel and tugged its top branches towards the ground to form a broad, low arch; yet, paradoxically, here too the impression left is of a kind of elegance. Towards the westward end of the wood, several young hornbeams that have sprung up underneath the canopy are beginning to bend towards the ground of their own accord, as if unable to carry the weight of their foliage. Near them, young yews stand under the beech and limes in small, dark, skulking enclaves, turning their backs on all around them.

When I first enter the wood one pleasant morning, I leave my equipment next to a beech 200 yards from the fence and continue on up the track (there is still a track of sorts here), confident that it will be perfectly safe there. The wood lies in the midst of a popular and heavily used stretch of country, and there are walkers and cyclists everywhere today: the shouts of canoeists on the Wye carry up clearly enough for me to be able to make out some of their words, and there is the constant chiggering of a generator from the camp site on the opposite bank. Here again, wilderness − quasi-wilderness − stands in close proximity to a scene of constant human activity, although in this case the wood does not seem to be much touched by it.

Today is a day of sunshine and showers, and the wind is making a sound somewhere between hissing and a dry rustle in the newly opened leaves of the young birches and beeches above my head. There are flies here − the first I have seen this season. But then, it is nearly May by now: the bluebells are just open, and the honey bees are working them. I pause to watch a delightful black beetle heaving its tank-like form over the curled surfaces of last year's leaves, then continue down past a rather half-hearted piece of fencing marked DANGER. Beyond it lies a broad promontory of the valley cliff, and the development of the big trees on it is particularly beautiful. I am determined to concentrate my efforts here, as I have done already in a number of the other woods, taking pictures over an area of no more than a few hundred square yards. Since photography is selection, it seems a good principle with landscape to pre-select a small patch of it and search out its essential qualities rather than always ranging widely and hoping that chance will solve the problem for you.

I follow the ghost of a path steeply downwards − it could be an animal track, but something about its shape suggests that human feet still use it − until I see that the cliff here is shifting outwards in a series of landslips, with some trees riding on each, until the moment when they must topple outwards towards the river. There is even a cave lower down, but it has bars across it. In 1988, Mark carved his name on a handsome beech not far away.

I clamber back up to the track and head back in the direction of my equipment. The thought occurs to me, dimly at first but then with increasing clarity, that if Mark could get down here in 1988, then there is no reason why entire armies of professional camera thieves should not have ignored and passed the PLEASE KEEP OUT notices during the past 20 minutes. The first curse of the photographer, as opposed to the writer, is that he must hump his tonnage of equipment with him; the second, and far worse, is that he must waste inordinate amounts of energy in fearing for its safety. Again I hear shouting and laughter from far below, this time from a group of ramblers crossing the pedestrian suspension bridge that spans the Wye. As they do so, the bridge makes a highly distinctive sound, rather like that of massively squealing bed springs. My little burst of paranoia causes me to make a classic mistake, and I take a wrong turning at an obscure fork in the track. In another two minutes' walking I have covered what I know is the right amount of ground, and so begin to scan the trees ahead for sight of my encampment. I do not see it, of course, and hurry on, sweating in my wellies, which in any case are far too warm for the day.

I see the boundary fence ahead, with a large group of walkers moving beyond it – thieves, quite clearly thieves – and my eyes bore into them now for some glimpse of a tripod criminally stuffed inside an anorak, a camera bag cunningly disguised inside a rucksack. Disappointed in this, I cross the fence and recognize the full stupidity of my mistake. I slide down the mud-chute that passes for a footpath here until I get back to my first point of entry. No one has touched my cameras. Of course no one has been into the wood.

The Forestry Commission is responsible for the management of a large proportion of the Forest of Dean, of which Lady Park Wood is one small part. Once a very rich ancient landscape, this forest has been horribly degraded, first by the establishment in the nineteenth century of large areas of poorly managed oak plantations, and second – and of course much more seriously – by the blanket planting of conifers under that same government body which, until 1984 at least, would have been more accurately known as the Conifer Commission. Since its change of policy, however, the Commission has had a rather more sympathetic attitude towards broad-leaved woods: currently it is preparing a new broadleaves policy, and thus plan of management, for the Forest of Dean as a whole. It is taken as a fundamental of this policy that, where they now survive, ancient woodland sites in the Forest should be left intact and encouraged to regenerate, drawing upon their own resources as much as possible. It would be good indeed if the Commission saw fit to roll back its carpets of conifers a few hundred yards from the margins of every such site, and thus allow it the chance to expand outwards a little. One experiment of precisely this kind is currently under way in Haugh Wood, Herefordshire. This wood is also on Forestry Commission land.

A thick snaking of ivy on a mature beech. Management would long ago have removed this.

Exposed roots in a landslip at the cliff edge.

Spring foliage in mid-afternoon sunlight.

A yew clings to a rock face, with holly close above it.

The steep drop to the Wye, looking eastwards.

The forest floor: precipitous, moss-coated rock and scatterings of broken timber.

Ty·Newydd Wood

LATE APRIL

A small section (one hectare approx.) of a larger woodland of sessile oak, typical of a series of
such woods to be found on steep valley slopes in the Cambrian Mountains. It lies adjacent to the
Royal Society for the Protection of Birds' Dinas Reserve (purchased in 1968), some ten miles to
the north of Llandovery, Dyfed. Like the reserve, the wood is notable for the variety of its fungi
in autumn, and is managed by the RSPB to the benefit of its pied flycatchers and wood warblers.

Stand on the farm·road bridge that spans the little River Gwenffrwd, or on the reservoir road over·
looking the Dinas woodlands and the Doethie Valley, and you might be forgiven for thinking that
both these river valleys were forested from one side to another, and that the few fields visible had
been cut only yesterday out of some great deciduous forest that continued for hundreds of miles
across the out·of·sight hilltops. Alas, it takes only the briefest glance at the map to divest oneself of
all such illusions. In reality, the stands of deciduous trees here are sited only on the steepest and least
accessible slopes, mostly in thin strips running along the bunched contours between the more gently
shaped upland and the valley floor. And what of that upland? At present – in the hills running
northwards from the Mynydd Mallaen to the great bog of Tregaron – somewhere between one half
and two·thirds of the land area remains as open moorland. The rest is under conifers.

It is a particularly pleasing experience to return to this little stretch of woodland. It was here,
some ten years ago, that I first began to take serious note of the kinds of beauty to be found in
'wilderness' landscapes. At the time I photographed the trees and the wonderful clusterings of fungi
beneath them using a toy tripod – it was light enough to carry on the back of a bicycle – with pre·
dictable results. The pictures would have been masterpieces, naturally, had any of them been quite
in focus. Now, in late April, I find myself climbing the well·trodden sheep tracks on this hill with
rather more robust equipment, and (in the most vivid contrast with my experience at Kingley Vale,
see p. 65) within a few minutes recognizing individual trees, combinations of tree·root and rock, and
even certain pads of moss. With the strangest sense of time in dislocation I begin re·framing ancient
compositions in today's light. There has been no great change here, as yet, and it is good to find that
some grotesques – especially the trees that have rooted themselves directly into the rock face –
stand here just as I remember them.

It is this phenomenon of rooting in rock which distinguishes Ty-Newydd Wood from the other oak woods on the same set of inclines. Elsewhere the slopes are gentler, and there are few, if any, miniature cliff faces for the trees to cling to. In Ty-Newydd, the oaks find and hold their ground in the rock in places where the sheep cannot reach them; and it is by observing this combination of randomness and resiliency that one may perhaps get closest to some notion of the appearance of the wildwood in these mountains. Here, the roots are to be seen on all sides bulging out of crevices: the lowest parts of some boles have flattened themselves to the rock, while others seem to flow sinuously out and back — though again the movement is rather too slow for human vision — like broad-girthed snakes. In one place, the top branches of the most pinched and light-starved little oak protrude from a split in the cliff that is no more than eighteen inches wide. In another, a slab of rock has separated itself from the cliff, fallen on to the steep slope below and slid down against the base of one of the bigger free-standing trees. As if to accommodate the stone's pressure, the tree's bark has grown out around it, shaping itself into a kind of oaken mouth, which now holds the rock fixed in its grip against the pull of gravity, in a kiss.

As a landscape, this woodland has three distinct layers or divisions. The lowest of these is the steep slope that runs directly up from the road asphalt to the cliffs. In effect this is now a series of grazing terraces, connected by a complex network of sheep trails; and if I do note one change in the wood since my first visit, it is that the mosses here appear to grow less thickly than they did under the pressure of grazing. The biggest of the oaks are in this band. Above this layer is a range of miniature cliffs, where the rocks are beautifully patterned by pale green lichens, and it is here that the most individual trees are to be found. Above the cliffs the slope gradually becomes gentler, the oaks are smaller and much more conventionally shaped, and birches grow in thick stands in competition with them. Above the birches there runs a fence — keeping the sheep out at the top of the wood, if not at the bottom — beyond which the moorland voids begin.

On my first morning here it is a day of 'sunshine and showers', so promised, which present themselves as something rather more akin to continuous rain coming in pulses of varying intensity. The cloud is carried on a steady gale, and tiny patches of blue fly by in its midst at thrice the speed of Concorde. As is my habit, I find a triangular 'cave' about three feet in height beneath the roots of one of my favourite subjects, and stash my equipment here; and, for the duration of the heaviest downpours, myself also. As I sit huddled beneath the rock, gazing through the trees to the road at my feet, I begin to wonder if there is not a case to be made for regarding the landscape photographer as some kind of hermit manqué, forever drawn back to wild and rocky landscapes with a camera clutched in one hand in place of skull or rosary. My retreat here is filled with a nest of dry oak leaves, decorated with a few strands of wool from previous occupants. The wind is coming from over the back of the hill, and the spot is completely sheltered. It is not difficult to imagine sleeping out here without a tent even on a gale-torn night, listening to the wind howling in the trees beyond and barely feeling a stir of air.

Half of the Welsh deciduous woods that survived into this century have disappeared since 1940, and what remains now occupies precisely 3 per cent of the country's land surface. Woodland studies carried out in the early 1980s showed that even this tiny area was threatened with obliteration, for the now familiar reason that much of it — 80 per cent of the Snowdonia woods, according to one study, 89 per cent of the woods in Dyfed, according to another — was failing to regenerate.

Overgrazing by livestock was identified as the most significant cause: Welsh farmers had been driven by one EEC incentive after another to increase their stocking levels on the hills, and had come to regard the woods as nothing more than convenient bad-weather shelters for the animals. This was before the initiation of the Coed Cymru scheme in 1985, which seeks to solve the problem by encouraging farmers (with the help of grant aid) to fence off their woods *and* manage them again for their timber. This scheme has been greatly helped along by the development of a new piece of equipment, the mobile diesel-powered sawmill. Nowadays, if he wishes, any farmer can hire such a saw, take it up to his woods in a Land Rover, and cut timber into saleable products on the woodland floor. A load of wood that would have been of little use to the timber merchant can be transformed *in situ* into goods such as fence posts and 'craft timber', tripling its value in the process, even when the hire of the saw has been taken into account.

During its first six years of operation, Coed Cymru succeeded in bringing between 10 and 15 per cent of Welsh broad-leaved woods back under management, and animals were fenced out of 1,574 hectares of previously open ancient woodlands. Certainly this is a move in the right direction, but it remains to be seen whether the balance can be shifted for the country's woodlands as a whole. The hill farmers are not going to abandon sheep as their main source of livelihood without a long fight, and it may be that pressure on the existing fragments could be eased just as effectively by using them selectively for grazing, on a 'rotational' basis (as noted elsewhere, this can be positively beneficial to conservation ends) and, at the same time, greatly increasing the area of each wood by planting saplings grown from seed from each site on adjacent land. There is no reason why the lovely hanging woods of the Gwenffrwd and Doethie Valleys should not be able to clamber a few hundred yards further outwards over the bare mountainsides, where soil conditions allow. There is no doubt that they would make the most wonderful change from conifer plantations.

Ty-Newydd wood exists at present in a 'parkland' state — a term one has come to understand to be a euphemism for 'grazed-out, dying on its feet'. The RSPB does now have plans to fence the wood at the road; but it will not exclude sheep entirely, as it aims to maintain it as an open-structured wood for the benefit of its bird population.

Trees rooted in rock. All are sessile oaks.

Seen from across the valley: Ty-Newydd Wood occupies the steep ground to the left of the road.

120

The textures of the mosses and lichens blur the expected distinction between the rock faces . . .

. . . and the trees that have colonized them.

An isolated ash tree on grazed limestone pavement at the wood's northern end.

Colt Park Wood

Late March

A National Nature Reserve of six and a half hectares, owned and managed by English Nature
and situated in the Craven Pennines above the Ribble Valley, some 12 miles north of Settle,
Yorkshire. An outstanding example of a rare woodland type, high-level ash wood on limestone
pavement. It harbours limestone grassland plants such as blue moor-grass and mossy saxifrage, as
well as baneberry, giant bellflower, globeflower and the rare mountain fern, spleenwort; it is also
notable for a vigorous growth of lichens and mosses.

For those with an interest in woodlands as landscape, there may be some use in this simple set of
equations: the more inhospitable the terrain, the more specialized the plant life to be found there.
The more specialized the plant life (and the longer it is left to do as it wishes in any one spot), the
more striking, bizarre and memorable the beauty that results.

Certainly this axiom applies without adjustment in the case of Colt Park Wood, where nature
has been at infinite pains to find a means of colonizing the seemingly impossible terrain of the
Carboniferous Great Scar limestone. Here under Ingleborough are wide, wind-blasted levels of
exposed rock which, during many millennia of water erosion, have been selectively sliced up into
something resembling a cockeyed aerial view of an American city, in which blocks of limestone
stand divided off from one another by a network of criss-crossing channels. Every third-year school
child knows that these are called 'clints and grikes', but none ever remembers which is which.

Colt Park Wood occupies the very edge of one such pavement expanse: the larger part of this
has a natural soil cover on which pastures have long been established. The wood runs north–south
in a thin strip above a 15-foot cliff which fringes the limestone on the east. With the addition of a
few well-placed bits of walling, this cliff is enough to keep the sheep out, while on the west the
wood is continuously walled, and it is to the presence of these barriers that it owes its survival.
Sceptical questions are now being asked about the long-established concept of climax woodland;
that is, the hypothetical final stage of evolution in a wood, in which all its plants exist in balance
with one another, and will continue to reproduce themselves indefinitely within this balance.
Nevertheless, it remains a useful concept, and can be applied with some imagination in a wood such
as Colt Park, which has been said to represent the climax woodland of the limestone pavements. If

the pavements had been left alone for long enough, they could all have looked something like this: that is the argument. As so often, the pavements' present-day nakedness is probably the end result of sheep grazing. Of course, it is also possible that early farmers actively cleared the woods for fuel. In any case, Colt Park Wood survives today by virtue of the unusually deep and dangerous grikes across which it is established. Farmers lost too many sheep to them, and so the wood was walled in. It is noteworthy, then, that this is a wood that has been able to regenerate itself without the theoretically beneficial effects even of moderate sheep grazing: it is, however, grazed by rabbits.

At Colt Park one enters a fragment − a tiny surviving splinter − of woodland that has unbroken continuity with the mountain wildwood of these parts. But no one should come here expecting to see looming depths and shade, or gigantic ancient trees. This is a high-level wood, and it is dominated by ash. One's first impression is of a scrubby little copse − what a farmer whom I once knew liked to call 'rubbish woodland' − in which even the biggest trees are not much more than 30 feet in height and many others are stunted, clinging on to life in a place where this is just possible. Some of them are fallen and growing prostrate, others stand even though they have long since died, with the bark dropping away from them patch by patch.

On my first day at Colt Park Wood, rain is in the offing. I enter the wood without equipment − by no means an unwise move, in this case − to do a little direct observation. Immediately I am across the stile, I find myself having to clamber with unusual care over mossed and slippery rocks, balancing my way from one foothold to another over grike after grike of indeterminate depth. In places, the floor of the wood is so carpeted by mosses that they have grown together across the grikes, making a perfect trap even for the wary. Other fissures are more open to view, and as I look down into them I wonder whether I will see any skeletons of the sheep that are said to be hidden there.

This is the kind of terrain in which, whenever you are moving, you find yourself spending 90 per cent of your energy just in covering the ground. Even direct observation is not easy, and after 20 minutes of fumbling and balancing I see with amazement that I have progressed no more than 300 yards. Up there beyond the wall is a barn, now converted into an office for English Nature, and I *know* that it is no more than spitting distance from my starting point. For just a moment I am able to picture what it might have been like to be stranded in the middle of one of Ingleborough's biggest pavements, in the distant days when it was under trees like this. As I am doing so, I lose my glove down a grike and have to hang over its damp, mossed sides with a stick in order to retrieve it. I make a clear mental note to drop no equipment here.

At a point some 400 yards down the length of the wood I find a junction, or intersection, of grikes that is wide enough to act as a clearing of sorts. Here I sit, on what is in effect a lower level of the woodland floor, eight feet down from the rest. It is very sheltered here − it has its own, more humid microclimate − and is lined with a rich humus in which both common and less common plants are able to take root. Next to me, the first corkscrewing twists of ransoms are just opening: National Nature Reserve or not, it is impossible to avoid nibbling at a few of them, and I see that the rabbits here share my lack of inhibition. A freezing north wind is raging in the canopy above me: occasionally an iced trickle of air gets down here, but otherwise it is quite still. Half a mile below, towards the valley bottom, a train clatters by along the Settle to Carlisle railway: like the warden's barn, from such a vantage (or disadvantage) point it seems simultaneously close and distant.

It is too early for the globeflower, unfortunately. But even so spring has already taken hold here. The snowdrops are finishing, the bluebells are emergent, the dog's mercury looks as though it will not be long before it starts to form flowers, and the ash trees above are just coming into bud. Seated so far below them, I am able to see how the trees have got their root-holds in the rock, though the phenomenon is well masked by mosses. Near by, the rock itself is draped in trailing copes of moss, whose complex texture resembles that of some beautiful and ancient fabric. Some trees are growing not from the surface of the clints but from the floor of the intersection: when I stand I can see others protruding from their respective grikes, their lower portions invisible. A robin flutters about me here, circling, halting, circling again, and as I move slowly on through the wood it follows me.

Even in woodland of this kind, it is still possible to manipulate the wilderness, should the need or the opportunity arise. Some young ashes have recently been established here by planting; and as the art of the conservator must always imitate nature, a few of these have been cunningly inserted into the grikes, and grow out from ledges there. This is a most unusual kind of rockery.

Ash trees rooted in rock, and growing up out of the grikes.

Beneath the wood, the sheep-grazed pavement is quite bare of trees.

Two further certainties of the wildwood: fallen trees and impassable terrain. But one detail of later civilisation is visible here.

Boulders next to the low limestone cliff above which the wood is sited.

Rock formation and ash tree, near the centre of the wood.

On the middle slopes of the wood: last light, rain.

Ariundle Oakwood

EARLY APRIL

A National Nature Reserve of 70 hectares owned by Scottish Natural Heritage, lying to the
north of the village of Strontian on Loch Sunart, Lochaber, on a granite bedrock. A good example
of a native Highland oak wood, and a remnant of the series of such woods that once dominated
on south-facing slopes and in the sheltered glens of this part of Scotland, most of which were
cleared for agriculture or grazed out by livestock. Of special interest for its communities of liverworts, lichens and mosses; in particular the 'Atlantic' species, to which the mild, humid climate of
the Western Highlands is especially well-suited.

I see Ariundle's oaken canopy first from a mile or so to the south, in the half-light of a drizzling
afternoon. At this distance, the wood seems to reproduce in itself both the colours and the dif-
fused, rounded forms of the heavy cloud that hangs above it, and shuts out all view of the mountains that lie beyond. This cloud is 'heavy', yet the wood itself seems light: it looks as if it might
at any minute slip free from its moorings on the mountainside and drift upwards into the damp air
with which it holds affinity. A little closer to me, directly to its west, there stands a dark, unre-
generate slab of Forestry Commission conifers: this gives way in a straight line, rising diagonally
across the hill slope, to the feathery milk-maroons and greys and the rounded clustering outlines
of the oaks' top branches.

Few things better underscore the beauty of the surviving native woodlands than a close juxtaposition of this kind. In effect, the new monoculture planting is devoid of qualities, good or bad: it is
no more than a landscape blackout, a dead zone. Beyond it, however, all is variety. Even in this
gloaming, each oak's fan of upper branches registers on the eye as slightly darker at the base and
paler at the tips, and no two such fans are quite alike in outline or detail. Where the pine plantation
reduces its flank of mountainside to the look and texture of a dark, tilted doormat, with all the
excitement that that implies, the oak wood beyond clings to its valley-side with the greatest delicacy
and subtlety, revealing each vagary in the land it occupies in the lines of its canopy.

Nor does it disappoint when one draws closer. I drive past the sign to a hand-knitted woollens
shop and café, and continue on along a freshly tarred road to the car park. The open land on both
sides of this road is thinly set about with dozens of mature 'parkland' trees, a sure indication of its
former status as woodland. It is a good long walk from car park to the wood proper, and I have

already negotiated a dozen lochs today to get here. By now, like all true addicts deprived of their release, I am fractious and tetchy, hungering for my first 'shot', through which I can perhaps begin to make some connection with this unknown place. I am conscious of the fact that the light is going – once again, two hours ahead of schedule – and hear a familiar inner tug-of-war between one voice which says 'Give it up – you've already missed it', and another, arguably wiser, which counters 'Carry on! You won't catch it in *this* light again.' As in other parts of the Scottish periphery, wych-elms have survived here: I photograph a particularly fine specimen standing next to the track, convinced that I now have at least one masterpiece in the can. It is an unfailing law that all such shots are unusable. Even so, the business of getting it is enough to break the ice: having fixed one tree through the lens in this way, it is always possible to go on more productively to others.

The main body of Ariundle Oakwood is sited on a steep and generally south-facing slope, climbing up from a forestry track to a hidden horizon. Very distinctly, it possesses the look of a once-managed wood that has been allowed to grow old quietly. In the company of other oak woods in the area, Ariundle was managed during the early nineteenth century for the production of charcoal for an iron furnace at Bonawe, and was subsequently used for grazing. Almost all the oaks here are mature and there is little evidence of new saplings, even inside the owners' first experimental enclosure where, as yet, birch saplings are dominant. Since 1975, the deer and sheep have been excluded from the wood as a whole by new fencing, but as things stand it is not at all difficult to imagine the work of the loggers here, or, on the upper slopes, the coppicing that has produced smaller, irregularly formed trees, many possessed of multiple trunks.

One of the first impressions I have in this wood is of an almost startling *cleanness*: the air (which is both clean-smelling and free from all kinds of machine noise), the rain, the little streams (though I do not drink from them) that wind their way around boulders and tree roots, all seem to possess this quality in abundance. It is two days before I discover that the wood is situated less than a mile from a former lead mine, whose abandoned rust-orange filter beds can still be seen decorating the mountainside above Strontian. Even so, Ariundle could not support its wonderful population of lungworts, lichens and mosses, were it not clean to a degree. As in the best of the humid western woods, the boulders here are completely hidden under springing pads of moss: at a distance many of them resemble gigantic hedgehogs curled up in a permanent state of self-defence. The oaks too wear green coats but these are more variably clad, revealing the occasional patch of moss-free bark as if in violation of some Calvinist law of decent dress. Yet even in these spaces, tiny ferns somehow cling on.

Lungworts were once known as lungs-of-the-oak and grow here amongst the bark mosses, thrusting themselves out from the lush vertical sward like little curled leaves of lightly baked Savoy cabbage, tinted in shades of green and sea-green blue. Above all, in Ariundle, the forest floor is a treasure-house of bryophytic growth: on one broken stick you may without any effort find half a dozen different fantasies of creation – crinkled and broad-fingered lungworts; sopping green and purple mosses; greenish-white encrustations of lichen resembling, and almost as hard as, corals; dangling bell-shapes of slime-moulds made, it seems, of nothing more than a slopping brown jelly. These distinctly resemble tiny jellyfish and, as I prod them like a child (as no doubt I should not), in a fascinated mix of attraction and repulsion, I discover that some are rather more cohesive than others: the sloppiest of all spread easily along the wood like badly set marmalade. On a grander scale, too, the greened branches and tangling branchlets of the oaks resemble undersea forms, especially in this rain-obscured half-light.

Nature knows no absolute boundaries, least of all those laid down by human beings. I learn later from my bed-and-breakfast lady in her 1960s bungalow in Strontian that during her career she has played host to many a student of bryophytes, each of them having been drawn inexorably to the region in the knowledge of what they might find growing here. Her last guests, a group of ardent lungwort hunters, spent several days ranging Ariundle and neighbouring oak woods in search of a particular rarity, whose name she did not recall. They did indeed trace it to one of the woods, in a very remote spot, but by this time they were severely weakened by the efforts of their search and much in need of the Water of Life. It was only on the morning of their departure that the shortest of the three noticed that this very lungwort was growing quite happily at the base of the scrub apple tree next to which they had parked their car in their hostess's garden.

One of the wood's small streams.

A section of a boundary dyke, with the moss that coats it fallen away.

Wet ground with birch, at the wood's western end. Plantation conifers begin just beyond this.

Branch formations, looking down towards the centre of the wood.

The oak trunks make a very fertile ground for liverworts and mosses. Almost 250 species have been counted in this wood.

Early primrose, in the new enclosure. There are not many primroses to be found nowadays on the open Scottish mountainsides.

144

Rassal Ashwood

Early April

A National Nature Reserve of 85 hectares on private land, managed by Scottish Natural Heritage, sited on an outcrop of Durness limestone near the head of Loch Kishorn, Wester Ross. The most northerly ash wood in Britain, and one of very few natural ash woods in Scotland, it is the remnant of a much larger and more mixed woodland in which ash was present for about 6,500 years. It has been grazed continuously by crofters' sheep for the past two centuries, but also before this: 'Rassal' is Old Norse for horse field. Notable for lichen growth and the presence of some rarer plants, such as dark red helleborine and melancholy thistle.

Rassal does not look much from the road. From below, all one can make out is a small and not very inviting cluster of dark treetops rising above an intervening slope on which large ashes grow individually or in groups. But this canopy bears a disorienting resemblance to the kinds of small woods to be seen on limestone in the Pennines, or in the Mendips: there is nothing else resembling it in these remote north-western parts, and that should be enough in itself to excite one's curiosity. I park in the car park – a piece of level ground mostly occupied by the crofter's feed troughs – and clamber up the slope to the first ramp of mature trees. Their trunks are spectacularly blotched by near-white lichens and dark green mosses into patterns resembling the coats of cheetahs or hyenas. Above them the trees grow together more thickly. At first, against the sun-filled western sky, their bare branches strike me as strange and slightly disturbing: they seem to claw at the air in a far more demonstrative way than their counterparts in England. Thirty yards up the hill I come face to face with that rarest of sights at any height on the open Scottish mountainside: a fully opened primrose. Wild strawberries are just coming into flower on the bank alongside it.

There is an icy north-easterly today, but – I have been lucky with the wind – the ash wood is almost completely sheltered from it by the mountain ridge beyond. I am already warm from humping tripod and cameras up the hill, and in the bright sunlight it looks for once as if dusk is a long way off. There is nothing more tempting to rest in the open than a warm, sheltered spot on a cold day; and, of course, one of the essential elements of the creative process in photographing landscape is now and then to lie down upon one's subject. I use my camera bag as a pillow of sorts, and lie back on a cushion of soft grasses overlying moss beneath one of the biggest ashes. Immediately I do so, the first impression of strangeness leaves me, and I begin to enter into a state of becalmed intimacy with the place, in which I am able to see it – and take pleasure in it – without further complications. Two of my lenses rest near by on the dry mosses that coat a broad fallen branch: this moss

145

is as resilient to the touch as the curled hairs of a poodle's coat. There is a pleasing paradox in this apparent dryness on such days in the western woods, since one knows that it is a dryness of the surface only. Beneath the outer layer the moss is wet just as, beneath the raft of parched grass on which I am lying, the plants, leaf-mould and soil are wet. Today this wetness does not touch me; but it would take no more than ten minutes of rain to top up every level here to saturation point.

The feeling of intimacy remains with me later as I work. In such a blissful and absolute solitude as this I can leave my equipment at any distance, just so long as I remember where I put it. Rotted ash branches collapse beneath my feet as I walk on, though I barely see them amongst the long grasses of the outer enclosure, and wherever I leave my camera bag, little pieces of twig somehow find their way down inside it. The place is so free from all sounds but those of nature that, when I step upon some brittle wood, the cracks are clearly audible as echoes amongst the surrounding trees. In this becalmed physical and mental state, with the sun shining evenly from behind the thinnest screen of cloud, no tree seems to clamour for attention; yet all are beautiful, growing only more so as the light thickens. The youngest and least lichened trees stand out a ghostly grey amongst the rest, and this colour in particular intensifies during the last two hours of daylight.

It seems likely that the whole of this north-west-facing mountainside was once covered with a deciduous woodland, of which Rassal was merely one element. A core of the earth was taken near Loch an Lòin, a mile and a half to the north, which shows a complex deciduous presence there in the distant past; while Coille Dhubh, to the north-east, remains today as a woodland of over a mile in width in which birch is dominant. Of course, if such a big deciduous wood could hold its ground here, then the same must be true over other parts of what is now the barest terrain. Many people love the openness of Britain's acid bog and heather uplands and can think of nothing better than a long hike across them. I confess that I am not amongst them. I cling to a stubborn belief that this love of the wide open upland may be nothing more than an unconscious reversal of a long-buried collective fear of the forest: the impenetrable forest, home of dangerous beasts and brigands, through which journeys were made only at continual risk of accident or death; the forest that threatened all emergent notions of civilization and must be cut away, subdued and tamed to make space for that bedrock activity of human survival in the modern world, farming.

For myself, and especially in Scotland, there is a terrible gloom in the sight of the mountains, those vast, naked, heather-blacked rock monsters with their icings of snow and their sheep – the only immediately noticeable occupants – casually asleep even in the middle of the trunk roads. There is variety in the mountains' forms, certainly. But unless one is a botanist, and prepared to rifle on one's knees amongst the heathers, there is no perceptible variety in their texture. I look up at them as I travel and see nothing other than the landscape that has followed on from a multi-generational act of devastation, an ecological wasteland. The initial losses in the Scottish forests were due to climatic change, which turned many mountain forests into peat bogs. But from the Dark Ages onwards – and it is chastening to remember this today – the forests were frequently destroyed not merely by felling and subsequent grazing out, but by the sheer brute idiocy of *burning*.

At Rassal Ashwood, however, we can see in embryonic form and on a minute scale the results of some of the first attempts made by naturalists to encourage indigenous woodland to reverse this process using its own resources. It is a fundamental of all ancient woodland sites which have not been overplanted that the trees which grow there now will be highly adapted to their site. Hence, if a wood is to regenerate on its own ground, let alone to begin to spread outwards across the surrounding emptiness, the seed must be derived from the trees that already grow there. As noted in other woods, grazing by deer, cattle, or (as here) by sheep can reach such a level that even where saplings do establish themselves, they do not grow because they are continually browsed out. The first experimental step at Rassal, in 1958, was to exclude the sheep from a central area by fencing. By 1992, the contrast between the appearance of the wood outside and inside this fence is both striking and exciting. Outside, the trees grow thinly, in a state that does slightly resemble the planted parkland of a great house: most are mature or over-mature, and many are scarred at the bases of their trunks where the sheep have gnawed away the bark. In early April, the woodland floor is a largely uninteresting mix of long-dead grasses and a little heather, with moss showing up only in patches. Inside the fence, however – after a long period in which there was little regeneration – there now stands a young ash woodland dense enough to be difficult to penetrate. New saplings have sprung up here from seed, but additionally there has been a burst of 'lammas growth', unusual in ash, whereby the roots of the older trees have sent up shoots in much the same way as does garden lilac. At ground level, mosses cover almost every surface, and from them grow bizarre grey lichens that bear a strong resemblance to curled fallen leaves.

Like most ancient woods, Rassal was shrinking, and as the trees died the genetic base died with them. Between 1949 and 1980, 1 per cent of the trees were lost, mostly by being blown over. Had this process continued alongside the same levels of grazing, in time the wood would have thinned out and disappeared from the landscape. What this single piece of fencing shows, as if in diagrammatic form, is that such a fate is by no means inevitable. A second enclosure (more precisely, perhaps, they are referred to by the specialists as 'exclosures') was erected in 1975, at the top end of the wood. Some experimental planting was also done here on open ground above the old wood boundary. Other such plantings have followed in the wider expanse of woodland. The process could be described as a kind of gardening with wilderness – manipulative in the extreme, certainly, yet at the same time breaking no code of scientific rectitude. It does no more than what is intended, allowing woods to regain a little of the ground that has been lost.

The last fencing job as Rassal Ashwood was completed only in 1991, and this fence surrounds the woodland as a whole, following the pattern at Ariundle (see p. 135). The crofter here remains committed to rearing sheep, and is thus paid compensation for his loss of grazing. However, it may well be that his animals will not be permanently excluded from the ash wood. Research into the effects of grazing on woodland sites is still at a formative stage, and in the case of the upland woods in particular it is an issue of central importance to their future. On the strength of established information, however, it is already known that a 'moderate' amount of grazing by herbivores can be positively beneficial to their ecological diversity; after all, the wildwood came naturally furnished with such beasts.

With this help from outside forces, then, Rassal Ashwood is now growing a tiny extra margin on the hillside to its east, and shows more than a fighting chance of recolonizing the land within its earlier boundaries. It remains to be seen whether, by hook or by crook, the wood might be encouraged over time to spread out further across the open mountainside. But for this to happen at all it would be necessary for the crofters who farm these hills to accept the validity of such a modified land use. They should accept it, when the conditions are right. We have far too many sheep: we have only one such ash wood.

The 'parkland' structure of the new enclosure has been maintained in this state until very recently by sheep grazing.

After the grazing has stopped: an unusual form of regeneration direct from the roots of an old tree, inside the Number Two enclosure.

Two of the largest ashes, standing and recently wind-thrown.

Dusk: the ash branches seen from the open hillside to the south-east.

An isolated outlier, unprotected by fences. The bare moorland was once covered in deciduous forest.

Dead or dying pines stand here undisturbed amongst the living.

Eilean Suibhainn

EARLY APRIL

The largest of a group of some 43 privately owned islands totalling 658 hectares in Loch Maree, Wester Ross, managed as a National Nature Reserve by Scottish Natural Heritage. It contains one of the best remaining fragments of native pine forest in western Scotland, possessing an unusual understorey of upright juniper scrub as well as some uncommon plants, such as marsh club-moss, royal fern and brown beak-sedge. Wholly covered by primary woodland, none of the islands shows much historical evidence of human disturbance or attempts at management. Some areas of peat have developed into Scandinavian-type 'bog forest'.

If, even today, the Loch Maree islands seem to possess something of the quality of an earthly paradise, then this is for no better reason than that human beings have let them be and, for the most part, do not go there now. This is another reserve to which access is *not* encouraged except for the purposes of study, and in any case, unless they are ferried across by the Warden, visitors would need their own boat to get there as well as a means of launching it from a largely inaccessible shoreline. Though Eilean Suibhainn is well under a mile from the trunk road skirting Loch Maree, this is a genuinely remote place, the largest of a series of detached fragments of the mainland on which the ancient pine forest still survives largely unaltered, repeating its place-specific cycle of generation, competition and death in what is probably much the same way as it has done here since the time when the glaciers last melted away northwards.

I approach the islands in the company of the Warden, who has generously volunteered to take me out at little more than a moment's notice. Time is against us: heavy rain – expected to last for several days – is fast approaching, and in any case there is not much more than three hours of daylight left. As the islands draw closer across the choppy grey waters of this inland loch, I have my first impression of their forest cover. There are some breaks in the trees but for the most part they come down to the water, lining the low cliffs and shores. Even at a distance the woods have a look about them that is entirely different from that of the other pine forest fragments which I have seen in Scotland. Since there is no wood management, the dead and dying trees stand or lie alongside the living, the blanched outlines of their lifeless branches and trunks boldly marked out against the sombre foliage of the rest. The slightest thinning of the cloud enables a moving pool of light to pass along the shores of Suibhainn, and for just a moment this white-dark contrast is intensified to the level of the theatrical. The island stands out now like some fantasy of wilderness beyond which rears the vast backdrop of the lochside mountains, no longer a remote grey but a startling reddish-purple. I did not read Arthur Ransome as a child: even so, with only an adult's impressions of his writing, I feel for a moment as if I am moving in his world, or a more northerly variant of it. I scrabble about urgently for cameras, forgetting in the heat of the moment that I am in a boat and that quarter-second exposures tend not to come out so well in such conditions. But now the pool of light has moved on. Isolated drops of rain begin to fall, the channel between the islands narrows, and we move slowly in towards a shallow strip of sand overhung by heather, my point of disembarkation.

In his book *The End of Nature* the American writer Bill McKibben argues persuasively, and with many brilliant pirouettings in the presentation and re-presentation of his case from different angles, that nature as we once knew it has already ended. By this he means that today there is no organic process, and of course no place on earth, which can confidently be said to exist *entirely* free from the deleterious influences of human activity. Thus, even when I have got myself to a place as remote and untouched as Eilean Suibhainn, McKibben would argue, I must still glance over its foliage for, say, possible damage as a result of acid rain; or, failing this, for other and less immediately obvious influences on plant life as a result of global warming or damage to the ozone layer; or, at least, must listen out for the sound of military jets. Above all, I must never *relax*. Even if I can detect no aberration with the naked eye – even if specialized study shows no serious deterioration, to date, from the earlier pattern of growth – I must always harbour suspicions that human-influenced change is present. *Should* these clouds be gathering now, in quite the way they do?

And the truth is, I do these things; I do them! The originality of McKibben's book lies in the fact that it gives the clearest imaginable definition to an entrenched attitude of mind that has grown up alongside 'green consciousness' as an inevitable concomitant of it. Yet distilled to its essence in this way, we can see quite clearly that it is a form of paranoia. Doubt is the order of the day: no matter what the individual circumstances of a place, no matter how beautiful, seemingly untouched or physically remote it may be, we cannot but bring to it our fears for its safety – indeed, the more beautiful it is, the more we must fear for it. Thus, McKibben argues, it is the *idea* of nature as something pure, inviolate, separate from us, that has ended. As the melancholy child 'Little Time' from Hardy's *Jude the Obscure* says, 'I should like the flowers very very much, if I didn't keep on thinking they'd be all withered in a few days!'

An air-sea rescue helicopter did chop past along the loch while I was on Eilean Suibhainn. And, of course, one does not forget about the ozone, although since more often than not the blue skies of Scotland seem to be lost behind several thousand feet of rain cloud, for the most part the memory is a faint one. All these things aside, however, I must still record the fact that rarely, if ever, have I experienced quite so strong a sense of coming into a place that is truly pristine – pristine without complication – as on the ribbon-shore of that little island. Standing there on the sand it seemed offensive even to leave footprints, let alone to push up through the tall lings, or allow my boots to sink into the sphagnum mosses, which are usually touched only by the hooves of deer (the deer can swim: their grazing poses one potential problem, even here). I visited more physically striking woodlands in compiling this book. But I think none of them will linger in the imagination longer, or more vividly, than the pine forest I saw on two short walks on this island, where all my work had to be completed in a matter of a couple of hours, in a steady race against the dying light, while the Warden went off in his boat to monitor the local population of black-throated divers, freezing in the lake wind as he did so.

This perception of the pristine, in this place, is two-layered. First, of course, it must lie in the knowledge that the forest has been allowed to get on with what it has done best here since time immemorial, with only the most superficial interference from man by way of felling or planting (there may have been a little of the latter on the most northerly of the islands), let alone through colonization (the sites of a holy well and chapel are to be found on Isle Maree; another of the islands possesses the ruins of a single shieling). One might almost be able to guess this from the appearance of the place. But second, and just as important, this perception arises from the knowledge that the process of regeneration can continue here just as it has always done — all other things being equal — since the islands now have the protection of the state; it negotiates with the landowners to keep them as they are and, with absolute justice I think, shuts the public out.

In the last analysis, of course, it would be foolish to deny that the longevity of these island forests can only be guaranteed (in so far as anything is guaranteeable, in any kind of world) by fundamental changes in the activities of human beings well beyond their mountain horizon. Even so, I do not feel that I have *sinned* in setting aside my creative paranoia for two hours and taking pleasure in what, by staying away from it, we preserve.

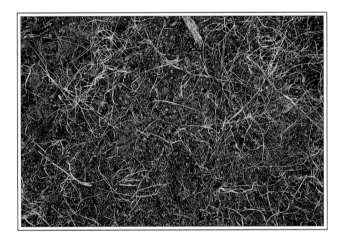

In the clean air, every exposed rock face has been beautifully patterned by the growth of lichens.

Young trees spring up at random among the vestigial forms of earlier generations.

An ageing pine. Beyond it are the cliffs of Creag Tharbh, which rear above the northern shore of the loch.

A freshwater lochan near the raised centre of the island, with the root-plate of a fallen tree.

Pines on a western shoreline. Upright juniper has also colonised this part of the island.

At the centre of the island.

Most of the reserve consists of open 'parkland' wood such as this, in which mature trees are not being replaced by saplings due to the pressure from deer grazing.

Glen Strathfarrar

MARCH INTO APRIL

The largest surviving fragment (2,250 hectares) of the 'Great Wood of Caledon' which, according to Ptolemy, once stretched the full width of the country from the Beauly Firth to the Argyll coast; sited along part of a glen some 25 miles west of Inverness. It is privately owned, but managed by Scottish Natural Heritage. Largely untouched in recent years by felling or underplanting, it harbours flowers such as the creeping lady's-tresses orchid and the rare one-flowered wintergreen.

The cover of the 1992 brochure of the Inverness, Loch Ness and Nairn Tourist Board shows one of those 'typical Scottish views', complete with familiar pair of gazing tourists, that are mandatory for such publications. The shot has been taken looking across a loch, but what dominates it is not water but the irregular, round-capped and beautiful growth of an indigenous pine wood on an island in the foreground. The far shore of the loch, too, is covered in what can only be a stretch of native pine forest: the bare mountain slopes above this are not at all important to the composition. Now, native woodlands today account for no more than 1 per cent of the land area in Scotland – 12,500 hectares in all – and 91 per cent of this total is to be found in the Highlands. Even so, it is quite clear that the Tourist Board considers that an image sufficiently definitive of the beauty of the Highlands to be worthy of the cover of its brochure should contain a prominent show of native woods. Implicitly, one is encouraged to think of the Scottish landscape in these terms, even if in practice it observes them for only 1 per cent of the time. Of course, one other perfectly reasonable inference to be drawn is that what holiday-makers habitually look for in the Scottish landscape is the kind of beauty to be found in those quite atypical places where fragments of the old woods still survive.

The road to Glen Strathfarrar is barred by a gate with nine padlocks, one of which may be opened for your car by the gatekeeper, if you are equipped with the appropriate permit. Beyond this, there are a good three miles of tarred road to traverse before you reach the obscure point of access to the Coille Gharbh, or Rough Wood, which lies at the heart of the Strathfarrar reserve. Along the first mile or so, the roadsides are dominated by oaks or birches whose branches have been thickened and clogged into bizarre green-woolled forms: they resemble skeletal arms adorned by badly knitted jumpers, but will seem strange only to those unfamiliar with the landscapes of eastern Scotland, where the phenomenon is quite common. As the valley narrows, its steep southern flank begins to dominate the view, and you find yourself looking up at a wooded landscape whose saturnine grandeur would have had immediate appeal to a romantic painter such as Caspar David Friedrich. Big pines grow on every slope here, as well as on some ledges on the rock faces, though the trees are never so closely spaced as to resemble a plantation; and, unlike the pinched little battery trees of every Scots pine plantation in the country, these have been able to grow to maturity with all the space they needed, so that their dark, rounded crowns dominate every vista. When I step from the car – though it is several degrees below zero today – for one moment I am reminded of the random growth of umbrella pines on the lower slopes of the Alpes Maritimes.

Again I begin my explorations without a camera, by entering one of the rocky clefts that run up northwards from the road. Everything in this cleft is beautiful: there is an infinity of patterns here, and the subtlest of colour combinations arising from what seem the most inspired juxtapositions of multicoloured mosses, lichens, bare wet rock faces, and dead and living tree bark. I walk alongside the stream, missing the path on higher ground. Mosses slide aside beneath me as I continue, my boot-prints remaining behind as deep holes in the sodden peat, and at every step my feet come out of the rich brown sludge with a noise like a sink-plunger. As I pause to note down an idea, isolated snowflakes bounce off or skitter down the surface of the pad. The iced wind is lovely, chilling my face so that my eyebrows ache, but failing to penetrate my many layers of clothes. I feel in balance with the weather, at least until I must climb the first hill with my cameras. The lightest dusting of snow drifts up the gully at me, then hardens into a few seconds of hail just before it clears. Beyond, I can hear the soft roaring of the River Farrar, and nothing more. Once again, it seems, this is a place where something approaching an absolute solitude may be experienced.

On the other side of the river, the Coille Gharbh is another place entirely. It too is intensely beautiful, though this is a more muted and elusive type of beauty. It is impossible not to think of painting when one is standing in such a landscape. Here the pink-barked pines rise out of a soft, even, rolling carpet of heathers, blaeberry, cowberry and sphagnum mosses, much as if they had been deliberately planted in pursuit of an ideal first established, say, in the backgrounds of an Altdorfer painting or the paradise fantasies of a Roelandt Savery. This is easily enough said, of course; but when one begins to take note of the complete absence of younger trees or saplings in the wood, it is no longer possible to maintain such a high aesthetic tone. Like Rassal Ashwood (see p. 145) before the fences went in, the Coille Gharbh is a 'parkland' wood, but its only architects are the beasts that graze it out and prevent its regeneration. In this case the grazing animals are the wild red deer, which are maintained and encouraged here by the owners of the glen specifically for the benefit of huntsmen.

Should you leave the Coille Gharbh and return to the road, as I did then, out of sheer bloody-minded curiosity, and continue on barely another half-mile up the glen, you would find yourself well placed to take some measure of the vast historical tragedy of the Scottish landscape. The road climbs a steep slope through what looks like a cutting, still overhung by woods, passes a dam, and then runs on alongside Loch Beannacharan through an upland valley of the most startling nakedness, in which such few natural fragments and even plantations as are to be seen stand islanded and lost. In Strathfarrar, just as elsewhere, the great pine forest was cleared piecemeal for farming until, in this case, the 1780s, when commercial timber extraction began there. Logs were floated down the Farrar to sawmills at Aigas, and shipped out from Beauly or used in the town's shipbuilding yards. The birch too was felled, for making bobbins. Sheep were introduced to the glen, but went into a swift decline by 1835. The deer 'forest' – it is difficult to think of a more knowing inversion of a word's true meaning – was established soon afterwards, a vacuous playground for the immensely rich into which, if you choose, you may buy yourself access today for the sum of £500 per gun per day, plus another £500 if by some miracle you happen to hit anything on the open hillside. In the reserve at dusk, meantime, the deer are so thick on the ground when I return to the gate that I am forced to drive down parts of the road at ten miles an hour or less. It is the easiest thing in the world to hit a deer with a car, in the forest: one does not so much as need to take aim.

Read with the slightest amount of insight, the signs along the Strathfarrar road also reflect the fundamental split in values that currently affects land use in this glen, as across much of the rest of the Highlands. One, erected on behalf of the Airdross and Aird Deer Management Group, warns of deer-stalking between mid-August and February. It does not *forbid* ramblers to walk on the high ground at this time, but it does warn that their presence 'can seriously disturb large areas for deer-stalking and can be dangerous'. Not far away from this board is the standard sign marking the western margin of the National Nature Reserve. This announces that the forest 'is being protected while maintaining the traditional red deer and hill farming uses'. It goes on to announce that 'few tree seedlings and saplings are able to survive the heavy grazing' which this maintenance of 'tradition', so-called, makes inevitable: an interesting contradiction.

The damage is inevitable only until the time when the forest is fenced against the hunters' moving targets, and the current chronic overgrazing is reduced to a level where it may be positively beneficial to conservation objectives. Now if this woodland is so very rare and, as a result, of such great importance to the nation, and if we know (as we do from one survey of the site) that 38 per cent of it has already been lost over a 40-year period, then I think we have a right to ask why it is not already fenced against the deer. One partial answer is that the reserve is not owned by Scottish Natural Heritage: under present and foreseeable financial conditions, land acquisition is regarded as a last resort in these matters. There is instead a management agreement, but the owners remain adamant that the reserve should be left open to the deer. Another factor is that, as at Rassal, some experimental fencing of a handful of selected patches of the forest has been carried out; although again, at the owners' request, none of these enclosures is greater than 20 hectares in extent.

I went into one of these, dating from 1980 and sited on a northerly slope, and found that – on what had been a hillside made virtually treeless by a fire – thousands of young pine saplings were now located well above the heather, naturally regenerated from seed that had presumably dropped there from the burned-out trees. It is not only the pines that grow back inside the fences, but the lichens and the mosses, which spring up amongst the heathers in a superabundance not to be found elsewhere. Especially beautiful are the sphagnums, which rise up here in rounded pads like engorged pincushions of soaking maroon-red velvet, six times the size of those outside.

There can be no doubt that under the right conditions fencing against deer will help the regeneration of natural pine forest. One argument against fencing in the whole of a wood such as the Coille Gharbh at once – and it is a good one – is that then the new growth would be all of the same age: the result would look something like a monoculture underplanting. But pocket fencing – such as has been begun here, staggered over time, and with the fences removed once the trees are well established after a period of 20 years – can begin the process of re-establishing the surviving indigenous pine woods as a series of habitats, containing all the stages between areas of young saplings and over-mature and dying trees that are seen in the abstract as a conservation ideal. Equally, as at Rassal and elsewhere, there is now the prospect of enabling such woods to spread out again at their fringes, fenced enclosure by fenced enclosure, across some at least of the bare slopes surrounding them. On average, pine seed can travel up to 100 yards from its source, but under the right wind conditions it can get much further. It must land in a spot where saplings can take root – it will fail, for example, where the moss is more than two feet deep. As at Rassal too, there is the prospect of planting saplings grown from seed taken from each site.

But for any of this to happen on any scale, the *owners* of the land must see the value of the idea. It is not enough, it seems, for the conservationists to offer them the prospect of free trees over time (some felling would be possible), or to proffer compensation fees for loss of grazing for the deer. First, the Highland laird (and nowadays the laird is just as often an absentee American, or Japanese, or a consortium, or management group, as an individual Scot) must accept that unless it is controlled overgrazing will eventually kill the forest fragments. He must also realize that the bare upland is not absolute, and has not always been like this: amazingly, some lairds believe that it has. Equally important, perhaps, for so long as this imperative remains in the Highlands, the laird may have to warm to the prospect of deer-stalking carried out in a genuine forest, rather than a 'deer forest', as has for example long been practised in Germany.

I do not think it is unrealistic, or grandiose, to suggest that the landscape of the Scottish Highlands could be utterly transformed in as little as 20 years if only the will – and one should add, the value system – were there. The incentives to such a change have never been greater than in the years since November 1989, when a new series of Forestry Commission grants was introduced for Scotland. The express aim of one block of these grants is to create 'new pine woods of natural character outwith existing pine woods', as well as to improve the management of the surviving native woodlands. Fencing is regarded as the normal method of excluding grazing animals, and grants are offered to pay for it; and natural regeneration is favoured over planting. Grants are also available for the creation, from scratch, of 'new pine woods which emulate the native pine wood eco-system' in all its variations from one part of Scotland to another, and a series of sensible guidelines is laid down towards the achievement of this end.

Today, therefore, the owners of pine forest fragments such as that in Glen Strathfarrar have an open invitation from government to indulge in a little gardening with wilderness of their own, with all due assistance and advice from Scottish Natural Heritage. Improved management of the sites – meaning the selective use of fencing over far larger areas than is currently seen at Strathfarrar – will be a major step in its own right, if it happens. But the opportunity now exists not merely to improve the quality of the surviving fragments, but to lay out the wilderness garden anew, in places where it has been obliterated for centuries: a facsimile paradise indeed, but far, far better than no paradise at all.

This prospect is particularly interesting when one considers the conifer plantation industry, which would surely benefit from any programme guaranteed to improve its justifiably dismal public image. The more opinion-sensitive plantationeers have already taken on board the notion that their cultivations can be done in such a way that they do at least possess a modicum of visual variety.

The mixing of species, and therefore of foliage colours, and the creation of irregular plantings using curves rather than straight lines, are both carried out now as a matter of course by some companies. In the last analysis, however, such efforts can do no more than make an industrial process look marginally less industrial. But is it too much to imagine that the big forestry companies might now themselves begin to make use of native pine wood grants? They would need to keep their facsimile native woods well clear of the plantations, since otherwise they might be colonized by the non-native conifers that are generally planted in the latter. This problem aside, however, I can think of no better way for Fountain Forestry or the Economic Forestry Group to begin to improve their public image than by dedicating, say, a minimum of 10 per cent of each new mountainside that they wish to afforest to the creation of a *permanent* native pine wood, complete with an access point and trail for those wishing to visit it. They would not need to look very hard to find models for such public-spirited works: Scottish Natural Heritage already has open to the public one of the most dramatic pine forest fragments (this is a genuine native forest, not an imitation!) on the slopes of Beinn Eighe, overlooking Loch Maree, incorporating both these features.

That is one approach. Another, less pragmatic but far purer of heart, has already been taken by the Trees For Life programme run by Alan Watson of the Findhorn Foundation. This aims to do no less than initiate the reconstruction of an entire uninterrupted wilderness landscape in those very naked mountains on both sides of Glen Strathfarrar. Alan Watson's team has so far planted some 8,000 trees, and fenced off for regeneration 105 hectares in the region, mostly on Forestry Commission land: not a huge area, but that is not the point. What matters most is the idea, one that won the programme the title of UK Conservation Project of the Year in 1991. Now that the Scottish Nature Conservancy Council has amalgamated with the Countryside Commission into Scottish Natural Heritage, it seems perfectly logical that the new organization should think seriously about emulating the English Countryside Commission's already operative scheme for a new 'national forest' in the Midlands. If it does so, then Trees For Life can supply the concept for a Scottish counterpart ready-made and working, and on a much grander scale.

The potential for the regeneration of the Scottish native forests has never been greater than it is today. The prospect might almost be said to be exciting, were not the future of such woods still almost entirely dependent upon the choices that are made, or not made, by landowners and afforesters. There is no imperative, no compulsion, in any of this, and as we have seen even on sites such as that in Glen Strathfarrar, the government's official conservation body can only act to conserve the surviving woods to the extent that the landowners allow it. No doubt, the Tourist Boards will continue to print pictures of native woodlands on their brochure covers; and tourists will continue to arrive in Scotland expecting to enjoy such landscapes without having to travel across miles of country before they find one. Whether it becomes any easier for them to do so or not is completely beyond the control of either party. I do not think it should be.

In one of the boulder-strewn gullies.

An isolated survivor at the head of a gully.

Outside the reserve. Here the native pine trees are now almost completely absent.

Glen Strathfarrar: detail of the root-plate of a fallen pine.

LOCATIONS AND ACCESS

These details supplement the information that is given at the head of each of the essays on the woods. OS references give the numbers of the relevant Ordnance Survey 1:50 000 series maps.

Black Tor Copse (OS 191)
This is an unfenced wood on Dartmoor common and hence, despite its rarity, it is not barred to public access. It is a good long walk up the valley of the West Okement River, which is itself a dead-end to walkers due to the presence of army ranges: the forbidden zone begins immediately to the south of the wood.

Lewesdon Hill (OS 193)
Open to the public via a network of public footpaths. However, parking close to the hill is difficult: one option would be to start from Broadwindsor.

Ashcombe Bottom Woods (OS 184)
Partially accessible by way of a public footpath running southwards along the combe between the Ox Drove and Tollard Royal.

Pinnick Wood (OS 195)
Fully accessible, as described earlier.

Kingley Vale (OS 197)
Accessible as described. There is a car park at the southern end of the bridleway, near the village of West Stoke.

Binswood (OS 186)
Open by way of a network of public footpaths. The Woodland Trust advises visitors to park on Green Street, the B-road to the north, and approach on foot along the bridleway that connects this road with Binswood.

Gutteridge's Wood (OS 175)
Open by way of a network of public footpaths.

Staverton Park (OS 156)
Accessible as described, only by way of a public footpath penetrating The Thicks from the B-road at its southern end. Parking is difficult here, however, and it may be better to walk to the park from the north, over a rather greater distance. The main area of old oaks has no right of way through it.

Lady Park Wood (OS 162)
No access, for the reasons given earlier. Alternative accessible deciduous woods in the same area include Little Doward and Bigsweir Wood (both Woodland Trust), and The Hudnalls (crossed by public paths). None has quite the same structure as Lady Park Wood, however.

Ty-Newydd Wood (OS 146)
No official access. The adjacent Dinas Reserve is circled by a public footpath, and there is also a nature trail through an oak wood some three miles up the Gwenffrwd/Nant Melyn Valleys, above Pont Rhyd-felin.

Colt Park Wood (OS 98)
Access not advised, except for the most determined, but is possible, by permit only. Apply to English Nature, North East Region Sub-Office, Thornborough Hall, Leyburn, North Yorkshire, DL8 5AB.

Ariundle Oakwood (OS 40)
Access possible by way of a woodland walk from a car park on Forestry Commission land (signposted from the public road).

Rassal Ashwood (OS 24)
Access to the outer enclosure is possible without a permit. A car park near the entrance gate is situated at the southern end of the reserve.

Eilean Suibhainn (OS 19)
No access, for the reasons given. There is no substitute possible, but the Beinn Eighe woodland trail is situated in the same area, to the south of Loch Maree and some two miles north-west of Kinlochewe. There is a car park next to the main road.

Glen Strathfarrar (OS 26)
There is free access to the glen for those approaching on foot or by bicycle. Those who wish to take cars in must obtain a permit from Scottish Natural Heritage, North West Region, Fraser Darling House, 9 Culduthel Road, Inverness IV2 4AG. A maximum of 25 cars is permitted on the road at any one time. Currently, the only approach to the Coille Gharbh (short of wading the river) is by way of a semi-derelict bridge, due for reconstruction. The next big glen southwards, Glen Affric, also has some fragments of old pine forest and is traversed by a public road.

Further information on British nature reserves can be obtained from:
England English Nature, Northminster House, Peterborough, PE1 1UA (tel. 0733-340345).
Wales Countryside Council for Wales, Plas Penrhos, Fford Penrhos, Bangor, LL57 2LQ (tel. 0248-370444).
Scotland Scottish Natural Heritage, 12 Hope Terrace, Edinburgh, EH9 2AS (tel. 031-447 4784).

Anyone wishing to visit a number of ancient woodlands should consider membership of the Woodland Trust, the body dedicated to woodland conservation. The Trust currently owns some 500 woods, of which 73 are designated as Sites of Special Scientific Interest. Its address is: Autumn Park, Grantham, Lincolnshire, NG31 6LL (tel. 0476-74297).

NOTES AND FURTHER READING

– RICHARD MABEY –

1. Oliver Rackham, *The History of the Countryside*, Dent, 1986. Rackham's books are essential reading for those interested in the history and ecology of ancient woodland. See especially: *Trees and Woodland in the British Landscape*, Dent, 1976 (new edition 1991); *Ancient Woodland*, Arnold, 1980.

2. An account of Bisham Wood is included in Peter Marren's lively and learned regional guide to the ancient woods of Britain: *The Wild Woods*, Nature Conservancy Council/David & Charles, 1992.

3. I have told the story of my 'community wood' in the Chilterns at greater length in *Home Country*, Century, 1990.

4. This account of woodland history is condensed from a treatment in: Richard Mabey, *The Common Ground*, Hutchinson, 1980 (new edition 1993).

5. Adam Watson, *Birds* (RSPB), 1991.

6. Roderick Nash, *Wilderness and the American Mind*, Yale University Press, 1967 (3rd edition 1982).

7. William Wordsworth, 'Description of the Scenery of the Lakes', 1822–3, in Wordsworth, *Selected Prose*, Penguin, 1988.

8. Bill McKibben, *The End of Nature*, Penguin, 1990.

9. Nan Fairbrother, *New Lives, New Landscapes*, The Architectural Press, 1970.

10. A useful history of changing social attitudes towards trees and woods is found in Part V of Keith Thomas's *Man and the Natural World*, Allen Lane, 1983.